Springer Series on

ATOMIC, OPTICAL, AND PLASMA PHYSICS 71

Springer Series on

ATOMIC, OPTICAL, AND PLASMA PHYSICS

The Springer Series on Atomic, Optical, and Plasma Physics covers in a comprehensive manner theory and experiment in the entire field of atoms and molecules and their interaction with electromagnetic radiation. Books in the series provide a rich source of new ideas and techniques with wide applications in fields such as chemistry, materials science, astrophysics, surface science, plasma technology, advanced optics, aeronomy, and engineering. Laser physics is a particular connecting theme that has provided much of the continuing impetus for new developments in the field. The purpose of the series is to cover the gap between standard undergraduate textbooks and the research literature with emphasis on the fundamental ideas, methods, techniques, and results in the field.

Please view available titles in *Springer Series on Atomic, Optical, and Plasma Physics* on series homepage http://www.springer.com/series/411

Jan Weiland

Stability and Transport in Magnetic Confinement Systems

With 51 Figures

Springer

Jan Weiland
Chalmers University of Technology
and EURATOM VR Association
Gothenburg, Sweden

ISSN 1615-5653
ISBN 978-1-4614-3742-0 ISBN 978-1-4614-3743-7 (eBook)
DOI 10.1007/978-1-4614-3743-7
Springer New York Heidelberg Dordrecht London

Library of Congress Control Number: 2012935694

Printed on acid-free paper

Springer is part of Springer Science+Business Media (www.springer.com)

Preface

This book presents the collective drift and MHD type modes in inhomogeneous plasmas from the point of view of two fluid and kinetic theory. It is based on a lecture series given at Chalmers University of Technology. The title of the lecture notes is Low frequency modes associated with drift motions in inhomogeneous plasmas. The level is undergraduate to graduate. Basic knowledge of electrodynamics and continuum mechanics is necessary and an elementary course in Plasma Physics is a desirable background for the student. The author is grateful to A. Zagorodny, I. Holod, V Zasenko, H. Nordman, A. Jarmén, R. Singh, P. Andersson, J.P. Mondt, H. Wilhelmsson, V.P. Pavlenko, H. Sanuki and C.S. Liu for many enlightening discussions, to G. Bateman, A. Kritz and P. Strand for collaboration on transport simulation, to my collaborators at JET, J.Christiansen, P. Mantica, V. Naulin, T.Tala, K. Crombe, E. Asp and L. Garzotti in modelling JET discharges and to H.G. Gustavsson for help with proofreading. Thanks are also due to the American Institute of Physics, the American Physical Society and Nuclear Fusion for allowing the use of several figures. Finally I extend my gratitude to my family, Wivan, Henrik and Helena for their continous encouragement and support.

Gothenburg, Sweden Jan Weiland

Contents

Chapter 1
Introduction

1.1 Principles for Confinement of Plasma by a Magnetic Field

The dominant purpose for confining plasma on earth is the achievement of nuclear fusion. The main path to achieve this is by using magnetic fields. Thus a scientific problem of major potential usefulness is the confinement of plasma by magnetic fields [1–26]. The necessity of plasma for fusion on earth is that the thermonuclear way of making nuclei collide by their thermal velocity is the only feasible way and this requires temperatures of above 100 million centigrades. At such temperatures matter is in its fourth state, the plasma state. A plasma can simply be described as an ionized gas where the charge of the particles makes magnetic confinement potentially possible. However, a magnetic field confines particles only in the perpendicular direction. Even in this direction the confinement is perfect only if the magnetic field is homogeneous and there are no other particles! Here, by confinement we mean that this single particle is not moving across the magnetic field on the average. When we have many particles, confinement means that we can maintain gradients in density and temperature. This, on the other hand, means that the system is not in thermodynamic equilibrium. For a plasma to be in thermodynamic equilibrium it must be homogeneous and have a Maxwellian velocity distribution. Thus a confined plasma will always be in a non-equilibrium state with different kinds of energy available to drive instabilities. An important aspect here is also that the magnetic field does not confine particles along itself. Attempts to cure this has been made by various types of mirror fields but the dominant and most successful method has been to bend the magnetic field into a torus. This introduces particle drifts due to the centrifugal force and inhomogeneity in the fieldstrength and these can drive instabilities. When the plasma density becomes sufficiently large, currents set up by perturbations in the plasma can significantly modify the external magnetic field.

J. Weiland, *Stability and Transport in Magnetic Confinement Systems*,
Springer Series on Atomic, Optical, and Plasma Physics 71,
DOI 10.1007/978-1-4614-3743-7_1,

Since thermodynamics always tries to take the plasma towards thermodynamical equilibrium, the currents set up by the plasma generally tends to change the magnetic field in such a way as to reduce confinement. When the current is mainly associated with particle motion we have current driven modes (Kink modes) and when the diamagnetic current is the main source we have pressure driven modes. The large scale versions of these are called *Magneto Hydro Dynamic* (MHD) modes. The instabilities of these mode, when fully developed, are so strong that a discharge is terminated on a short time scale (disruption) and thus, the system has to be designed in order to avoid these. Under operation MHD modes put the limit to pressure and current thus defining *Operational limits*. When the most dangerous MHD modes are stable we usually still have fairly large transport due to turbulence. This is something we can live with and that has been taken into account in the present ITER design. The turbulence is caused by small scale instabilities (microinstabilities) associated with drift motions in the plasma. The drift motions, in turn, are caused by the inhomogeneities in density and temperature thus closing our picture of relaxation in nonequilibrium systems. The corresponding eigenmodes are called *Drift waves*. While geometry is very important for the large scale MHD modes, drift modes can usually be described by the WKB approximation. Thus although geometry is sometimes important also for drift waves, the physics description (fluid, kinetic) is usually more important. Another important aspect is that transport is an irreversible type of motion and it requires irreversible properties of the generating equations. Since instability is the very source of the turbulence, the growth rate, as a part of the eigen frequency, also plays a very important role for transport. It thus causes the phase shift between potential and density or temperature (for $E \times B$ driven transport) which is needed for transport. The real eigenfrequency, on the other hand, describes periodic, reversible, behavior that reduces transport. This is why the dominant instabilities for transport are low frequency modes. The most important drift waves typically have real frequencies about 2 orders of magnitude below the ion cyclotron frequency.

In the present work we shall consider both macroscopic MHD modes and small scale drift type modes. Since we need the more detailed two fluid and kinetic descriptions for the drift-type modes we shall also use these for MHD-type modes. This allows us to see the connections and to make the transition between MHD and drift-type modes. We will, however, briefly discuss also the one fluid equations. Concerning the two fluid and kinetic approaches, we shall discuss them in rather much detail, discussing conditions for using two fluid equations. In particular we will give three different derivations for the lowest order Finite Larmor Radius (FLR) effect. Concerning wave particle resonances linear and nonlinear theory may give very different results since the nonlinear resonances have a tendency to counteract the linear ones. Here we expect the sources in velocity space to play a crucial role. We may compare this with the situation in real space where a background gradient is necessary for transport and we need a source to maintain a background gradient on a long timescale. A special section has been devoted to advanced fluid closures in toroidal systems.

Since the magnetic field confines a plasma in only two dimensions, the method of treating the problem with the third dimension is obviously very important. In a tokamak the toroidal curvature represents the third dimension. This means that the toroidal curvature is fundamental for the confinement. Its main obvious consequences are the presence of curvature driven modes and trapped particles. Since curvature is driving instabilities only on the outside of the torus, curvature also leads to eigenmodes that are trapped on the outside. These are generally called *Ballooning modes*. We note, however, that the term "ballooning mode" was originally introduced for the MHD ballooning mode and this meaning is sometimes still assumed to be understood.

Although the effects of toroidicity mentioned above have been known and studied for a long time, it is only since the end of the 1980s that strong efforts have been made to include them fully in calculations of tokamak transport. The main assumption to be removed from previous calculations of transport is that the diamagnetic drift, due to the pressure gradient, dominates over the magnetic drift which is due to toroidal (around the torus the long way) curvature and the closely related radial variation of the magnitude of the magnetic field. When this assumption is removed, a completely new regime of transport is introduced. This regime usually persists in the inner 80% of the small radius in a tokamak. For shots with highly peaked pressure profiles, such as supershots on TFTR, this regime is somewhat smaller while it is larger for shots with broad density profiles such as usual H-modes. In the new regime, transport coefficients tend to grow with radius which is in agreement with experiment and which was previously a main problem for drift wave models. In this regime the mode frequency is comparable to the magnetic drift frequency and this causes a problem with the conventional fluid closure, i.e. it requires advanced fluid models or interpretations. This will be discussed in the section on fluid closure. Since kinetically nonlinear effects are required close to resonances, the only alternative is to use a fully nonlinear gyrokinetic code. Although much progress has been made in that field, nonlinear gyrokinetic codes are still too time consuming to be run as transport codes only by themselves. Thus some combined system of transport code with continuous advancement of nonlinear kinetic transport coefficients in time is needed. Concerning advanced fluid models, these generally make use of several moments in the fluid hierarchy, making the closure at a level where remaining kinetic effects can be treated by some simplification. The energy equation is generally kept with its time dependence thus making a continuous transition between adiabatic and isothermal states possible. While Landau-fluid models here introduce linear dissipative kinetic resonances, the fluid model in Chap. 6 just keeps the diamagnetic energy flow, arguing that we do not have sources for higher order moments unless we have a heating source that is close to resonance with the drift waves. This is not the case for drift waves if we consider usual Neutral Beam (NB) or cyclotron resonance heating. Such a fluid model is here called reactive since the closure does not involve dissipation. The situation is, of course, completely different for fast particle modes (Chap. 8). A more advanced approach is to introduce a nonlinear frequency shift in the plasma dispersion function. This turned out to be quite successful for the case of only three modes but is algebraically chumbersam and has not been extended to a larger number of modes.

The significance of the ratio of magnetic to diamagnetic drift as the main toroidal effect for transport is shown by the fact that it is the largest one (this ratio goes to infinity at the axis), and that it enters dynamically through the pressure gradient. Such dynamics is important in transitions between different confinement states. It is also important to note that in the new regime mentioned above the density length scale drops out of the stability condition giving a condition that depends only on temperature and magnetic field length scales.

1.2 Energy Balance in a Fusion Reactor

Of course the theory developed for anomalous transport, as outlined in the previous section, aims at determining the confinement time in a reactor by first principles methods. In reality, the confinement times of new tokamaks have so far always been predicted by empirical methods on the design phase. The performance of both the previous (large) and the present ITER designs have, however, also been predicted by first principles methods. We will here start by deriving the condition on the confinement time required for energy balance or ignition in a fusion reactor. The time derivative of the energy density in a plasma depends on incoming and outgoing energy flows as:

$$\frac{\partial W}{\partial t} = P_\alpha + P_{in} - P_s - P_v \tag{1.1}$$

Where

$$P_\alpha = \frac{1}{4} n^2 E_\alpha {<}\sigma v{>}$$

$$P_v = \frac{3nT}{\tau_E}$$

$$P_s = 3.4 \cdot 10^{-15} {n_e}^2 {Z_{eff}}^2 \sqrt{T_e}$$

$$n_e = \sum n_j Z_j$$

$$Z_{eff} = \frac{1}{n_e} \sum_i n_i {z_i}^2$$

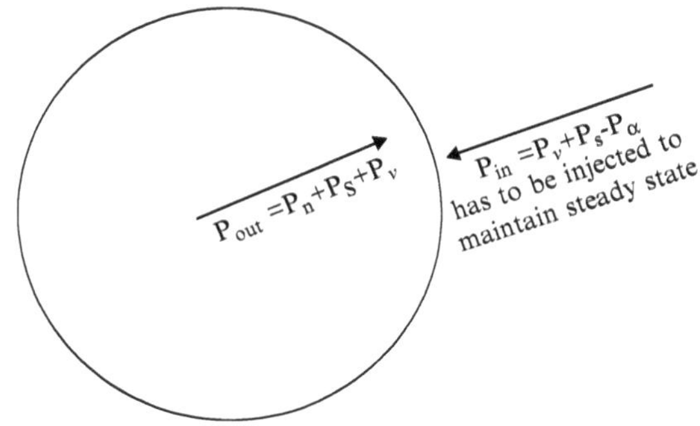

Fig. 1.1 Outgoing and ingoing powers in a fusion reactor

The outgoing and ingoing energy fluxes are (Fig. 1.1):

$$P_{ut} = P_n + P_s + \frac{3nT}{\tau_E}$$

$$P_{in} = \frac{3nT}{\tau_E} + P_s - P_\alpha$$

Now, the electric power obtained from the energy outflux is $\eta_t P_{ut}$ and the electric power required for heating is P_{in}/η_h. Thus the condition for a driven reactor to produce net energy is:

$$\eta_t \eta_h P_{ut} \geq P_{in} \tag{1.2}$$

This leads to the condition:

$$n\tau_E > \frac{3n^2 T(1 - \eta_t \eta_h)}{P_\alpha + \eta_t \eta_h P_n - (1 - \eta_t \eta_h) P_s} \tag{1.3}$$

Using the fusion cross section for the DT reaction in Fig. 1.2 we get the condition:

$$n\tau_E T > 10^{21}\,\mathrm{m^{-3}\,s\,KeV} \tag{1.4}$$

where the product $n\tau_E T$ is generally called the fusion product. Equation 1.4 is the condition for power *breakeven* Lawson criterion. If we require that we do not have to heat from outside we get the condition

$$n\tau_E T > 5 10^{21}\mathrm{m^{-3} s\ KeV} \tag{1.5}$$

which is the condition for *Ignition* (Fig. 1.3).

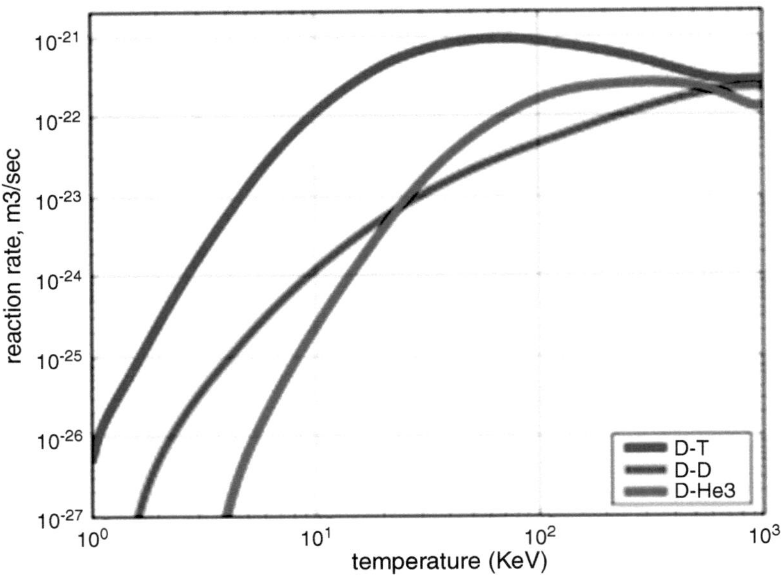

Fig. 1.2 Fusion cross sections for the DT, DD and DHe^3 reactions

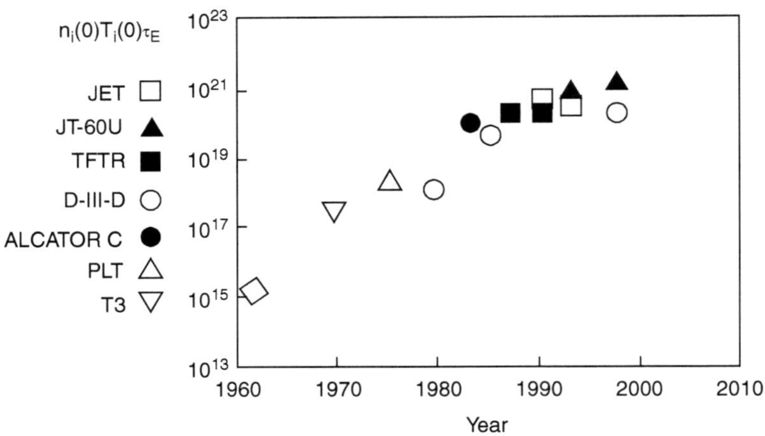

Fig. 1.3 The development of achieved fusion product over the years

Since experiments show that (1.3) cannot be fulfilled when MHD ballooning modes are unstable, i.e. when (3.30) is not fulfilled we also obtain the condition

$$\tau_E > \frac{Rq^2}{a} \frac{10^{21}}{B^2/2\mu_0} s \tag{1.6}$$

which now is a condition on τ_E alone. Here a is the minor radius, R is the major radius, q is the safety factor (given by the ratio of toroidal angle to poloidal angle variations as we move along a field line, Chap. 6) and B is the toroidal magnetic field. As an example this limit takes the value 4 s for JET with B = 3.5 T and I = 4.8 MA. The β limit (3.30) is due to MHD ballooning modes. When we also include the stability limit due to kink modes (Chaps. 3 and 6) the maximum average beta is given by the Troyon limit (1.8), [4].

1.3 Magnetohydrodynamic Stability

As mentioned above, MHD stability depends on both pressure and current driven modes. A combination of such modes enter in the Troyon limit for the maximum ratio of plasma and magnetic field pressures denoted β.

$$\beta = \frac{nT}{\frac{1}{2\mu_0}B^2} \tag{1.7}$$

The Troyon limit is:

$$<\beta> \leq g\frac{I}{aB_\phi} \tag{1.8}$$

Where < > indicates volume average, I is the toroidal current, B is the toroidal magnetic field and g is a numerical factor between 2.8 and 4.4 which depends on elongation and ellipticity. The Troyon limit, which concerns only ideal modes, gives maximum average beta around 5%. However, long term confinement is limited to average beta around 2.5% due to resistive modes. If we look at ballooning and kink stability separately a usual condition for Ballooning modes is

$$\beta \leq \frac{a}{Rq^2}$$

where a is the small radius, R is the large radius and q is a measure of the pinch angle of the magnetic field as will be given in detail later. It decreases with current. Kink stability is on the other hand given by

$$q \geq \frac{m}{n}$$

Where m is the poloidal (around the cross section) modenumber and n is the toroidal modenumber. Thus we see that current destabilizes kink modes but stabilizes ballooning modes. Thus the Troyon limit is a compromise between these conditions.

Another constraint of MHD type is the Greenwald limit.

$$n_G = I_{MA}/(\pi a^2) \tag{1.9}$$

This is a limitation of the density where I_{ma} is the plasma current in MA. This limit applies only at the edge and seems to be related to resistivity and radiation.

1.4 Transport

When the plasma β is in the MHD stable regime, the confinement is on good grounds believed to be limited by turbulent transport. This transport has in a rather wide regime for ohmically heated plasmas been observed to follow the so called Alcator scaling [5].

$$\tau_E \approx 3.8 \times 10^{21} na^2 \tag{1.10}$$

This scaling has recently been recovered theoretically as due to the dissipative trapped electron drift mode [6] or the microtearing mode [7]. These modes are both driven by temperature gradients and the density dependence comes from a dependence on resistivity. A further discussion of the modes is contained in Chap. 6. When the density reaches high enough values the Alcator scaling is saturated and a region where τ_E is almost independent of n enters. The energy transport in this region is believed to be due to a drift wave driven by ion compressibility effects in combination with ion temperature gradients. This is the η_i mode [8] ($\eta = d \ln T/d \ln n$) As it turns out both the trapped electron mode and the η_i mode are in the experiments typically not far from marginal stability [9] in the so called confinement region, of the plasma. This is an indication that these modes actually govern the temperature profiles giving rise to the so called *profile resilience* [10]. This is a typical feature observed in tokamak plasmas where the temperature profiles are virtually independent of the power deposition profile by neutral beam or radio frequency heating. A close relation between modes driven by temperature gradients and energy transport is also expected from thermodynamic points of view since a temperature gradient means a deviation from thermodynamic equilibrium and since an energy transport would tend to equilibrate the system. In connection with auxiliary (non Ohmic) heating a degradation in confinement (L mode) has been observed. In 1982, a new type of confinement mode, the H mode, was discovered on the ASDEX tokamak in Garching [11]. In this regime the confinement time is a factor 2–3 larger than in L mode. The confinement time does, however, degrade with power also in H-mode. The transport research has, over the years, been conducted both by empirical and first principles methods. Empirically one has derived scalings of confinement time with various characteristic parameters of the experiments. A very fruitful theoretical approach is to derive constraints on these scalings for consistency with the basic physics description [12, 13] (see the next section).

1.5 Scaling Laws for Confinement of Plasma in Toroidal Systems

So far scaling laws have been used to predict the confinement in all new tokamaks.

The foundation for these is that it can be shown that the thermal conductivity in a toroidal system has the general form:

$$\chi = D_B\left(\frac{\rho}{a}\right)^{\alpha} f\left(\varepsilon_n, \beta, q, \eta, \frac{L_n}{a}, \frac{a}{R}, \ldots\right)$$

Where ρ is the gyroradius, $\varepsilon_n = 2L_n/L_B$, L_n and L_B are length scales of density and magnetic field according to the general definition $L_j = -j/(dj/dr)$ and $D_B = T/(eB)$ is the Bohm diffusivity. The coefficient α is a coefficient that characterizes the transport. Thus while all parameters that are arguments of f are dimensionless, independent of system size, the factor in front will determine the scaling with system size and magnetic field. There are, in particular two types of transport that have been observed and discussed in the literature. They are Bohm diffusion corresponding to $\alpha = 0$ and Gyro-Bohm corresponding to $\alpha = 1$. In Bohm diffusion, mainly observed near the edge, transport is due to rather global modes that depend on the system size while Gyro – Bohm diffusion, observed in the core, is due to local modes that depend more on the gyroradius. Gyro Bohm transport gives a more optimistic extrapolation to larger systems with stronger magnetic fields.

A widely used scaling of the confinement time in H mode is IPB98(y,1) [15]:

$$\tau_E = 0.0562 \cdot I_p{}^{0.98} n^{0.41} B_T{}^{0.15} R^{1.97} \varepsilon^{0.58} \kappa^{0.78} P^{-0.69} M^{0.19} \tag{1.11}$$

1.6 The Standpoint of Fusion Research Today

As most of our readers know, ITER (The way in Latin) is now being built in Cadarache, France. The design is essentially that of ITER Feat from 2001 but after ITER was approved in 2006, a design review was conducted. ITER will be a tokamak with 6 m large radius and 2 m horizontal minor radius with elongation 1.6. Its magnetic field will be 5.3 T and the fusion Q 10 or more depending on plasma current. With Q = 10 the fusion power will be around 500 MW. It has been constructed from empirical scaling laws (like the IPB98(y,1)). However, dimensionless scalings from the performance of today's large tokamaks, like JET, JT60-U, DIII-D and Asdex UpGrade have also been used. The ITER design is conservative, i.e. new improvements that do not have sufficient reproducibility have not been included in the design. Examples are internal transport barriers, particle and momentum pinches and the Hybrid mode.

References

1. J.D. Lawson Proceedings of the Physical Society, London B**70**, 6 (1957).
2. J.R. McNally, Jr, Nuclear Fusion **17**, 1273 (1977).
3. W.M. Stacey, Jr, "Fusion Plasma Analysis", Wiley, New York 1981.
4. F. Troyon, R. Gruber, H. Saurenmann, S. Semenzato and S. Succi, Plasma Phys. Control. Fusion **26**, 209 (1984).
5. M. Gaudreau, A. Gondhalekar, M.H. Hughes et.al. Phys. Rev. Lett. **39**, 1266 (1977).
6. B.B. Kadomtsev and O.P. Pogutse, Soviet Physics Doclady **14**, 470 (1969).
7. J.F. Drake, N.T. Gladd, C.S. Liu and C.L. Chang, Phys. Rev. Lett. **44**, 994 (1980).
8. L.I. Rudakov and R.Z. Sagdeev, Sov. Phys. Doklady **6,** 415 (1961).
9. W.M. Manheimer and T.M. Antonsen, Phys. Fluids **22**, 957 (1979).
10. B. Coppi, Comments Plasma Phys. Controll. Fusion **5,** 261 (1980).
11. F. Wagner, G. Becker, K. Behringer et. al. Phys. Rev. Lett. **53**, 1453 (1982).
12. B.B. Kadomtsev, Sov. J. Plasma Phys **1**, 295 (1975).
13. J.W. Connor and J.B. Taylor, Nuclear Fusion **17**, 1047 (1977).
14. O. Kardaun, F. Ryter, U. Stroth et. al. "*ITER: Analysis of the H-mode confinement and threshold databases*", in Proc. 14th Int. Conf. Plasma Physics and Controlled Nuclear Fusion Research, vol **3**, Vienna, IAEA 1992 p.251.
15. ITER Physics Basis Editors, ITER Physics Expert Groups, ITER Joint Central Team and Physics Integration Unit, *ITER Physics Basis*, Nucl. Fusion **39**, 2137–2657 December (1999).
16. J. Wesson, *Tokamaks*, 3rd edition Clarendon Press, Oxford UK, 2003.
17. V. Mukhovatov, M. Shimada, A. Chudnovsky, A.E. Costley, Y. Gribov, G.Federici, O. Kardaun, A.S. Kukushkin, A. Polevoi, V.D. Pustivitov, Y. Shimomura, T. Sugie, M. Sugihara and G. Vayakis, *Overview of Physics Basis for ITER*, PPCF **45**, A235 (2003).
18. V. Mukhovatov, Y. Shimomura, A. Polevoi, M. Shimada, M. Sugihara, G. Bateman, J.G. Cordey, O. Kardaun, G. Pereverzev, I. Voitsekovich, J. Weiland, O. Zolotukhin, A. Chudnovsky, A.H. Kritz, A. Kukushkin, T.Onjun, A. Pankin and F.W. Perkins, *Comparison of ITER performance predicted by semi-empirical and theory based transport models,* Nucl. Fusion **43**, 942 (2003).
19. M. Shimada, V. Mukhovatov, G. Federici, Y. Gribov, A. Kukushkin, Y. Murakami, A. Polevoi, V. Pustovitov, S. Sengoku and M. Sugihara, *Performance of ITER as a burning plasma experiment*, Nucl. Fusion **44**, 350 (2004).
20. Editors of 'Progress in ITER Physics Basis', ITPA Topical Group Chairs, Cochairs and Chapter Coordinators, *Progress in ITER Physics Basis*, Nucl. Fusion **47**, S1 –S413, (2007).
21. G. McCracken and P. Stott, *Fusion-the Energy of the Universe*, Elsevier Academic Press (Burlington and San Diego) ISBN 012481851X, 2005.
22. V.V. Parail, *Energy and particle transport in plasmas with transport barriers*, PPCF **44**, A63 (2002).
23. F. Wagner and U. Stroth, *Plasma Phys. Control. Fusion* **35**, 1321–1371 (1993).
24. J.E. Rice, E.S. Marmar, F. Bombarda and L. Qu, Nucl. Fusion **37**, 421 (1997).
25. C.E. Kessel, G. Giruzzi,A.C.C. Sips et.al., *Simulation of the hybrid and steady state advanced operating modes in ITER*, Nucl. Fusion **47**, 1274 (2007).
26. D.J. Ward, *The Physics of Demo*, Invited paper 37th EPS conference, Dublin 2010, PPCF **52**, 124033 (2010).

Chapter 2
Different Ways of Describing Plasma Dynamics

2.1 General Particle Description, Liouville and Klimontovich Equations

In order to realize which approximations that are made in the descriptions of plasmas that we generally use [1–25], it is instructive to start from the most general description which includes all individual particles and their correlations in the six dimensional phase space (**r**,**v**). I the absence of particle sources or sinks we must have a continuity equation for the delta function density N:

$$N(X,t) = \sum_{i=1}^{N} (X - X_i(t)) \qquad X = (\mathbf{r}, \mathbf{v}) \tag{2.1}$$

$$\frac{\partial}{\partial t}N + \sum_i \frac{\partial}{\partial r_i}\left(N\frac{\partial r_i}{\partial t}\right) + \sum_i \frac{\partial}{\partial \mathrm{v}_i}\left(N\frac{\partial \mathrm{v}_i}{\partial t}\right) = 0, \tag{2.2}$$

Since we have included all particles this system conserves energy if we ignore radiation.

Thus there must be a Hamiltonian for the system and we use the Hamilton equations:

$$\frac{\partial r_i}{\partial t} = \frac{\partial H}{\partial \mathrm{v_i}} \quad ; \quad \frac{\partial \mathrm{v}_i}{\partial t} = -\frac{\partial H}{\partial \mathrm{r_i}} \tag{2.3a}$$

Leading to the form

$$\frac{\partial}{\partial t}N + \sum_i \frac{\partial r_i}{\partial t}\frac{\partial N}{\partial \mathrm{r_i}} + \sum_i \frac{\partial \mathrm{v}_i}{\partial t}\frac{\partial N}{\partial \mathrm{v_i}} = 0, \tag{2.3b}$$

J. Weiland, *Stability and Transport in Magnetic Confinement Systems*,
Springer Series on Atomic, Optical, and Plasma Physics 71,
DOI 10.1007/978-1-4614-3743-7_2, © Springer Science+Business Media New York 2012

Using acceleration due to the Lorenz force we then get

$$\left\{\frac{\partial}{\partial t}+\vec{v}\cdot\frac{\partial}{\partial\vec{r}}+\frac{e}{m}\vec{E}\cdot\frac{\partial}{\partial\vec{v}}+\Omega(\vec{v}\times\hat{e}_{||})\cdot\frac{\partial}{\partial\vec{v}}\right\}N(X,t)=0 \tag{2.4}$$

Where we wrote $\frac{e\mathbf{B}_i}{m}=\Omega_c\hat{\mathbf{e}}_{||}$. This can easily be generalized to the electromagnetic case. Equation 2.4 is written as a conservation along orbits in phase space. i.e.

$$\frac{DN}{Dt}=0$$

Where

$$\frac{D}{Dt}=\frac{\partial}{\partial t}+\vec{v}\cdot\frac{\partial}{\partial\vec{r}}+\frac{e}{m}\vec{E}\cdot\frac{\partial}{\partial\vec{v}}+\Omega(\vec{v}\times\hat{e}_{||})\cdot\frac{\partial}{\partial\vec{v}}$$

Is the total operator in (2.4). Equation 2.4 is the Liouville or Klimontovich equation. Since *N(X,t)*, as given by (2.1), contains the simultaneous location of all particles in phase space, it can be considered as a probability density in phase space. It gives the probability of finding a particle in the location (**r**,**v**) given the simultaneous locations ($\mathbf{r_i}$,$\mathbf{v_i}$) of all the other particles. This is an enormous amount of information which is usually not needed. This information can be reduced by integrating over the positions of several other particles giving an hierarchy of distribution functions (the BBGKY hierarchy) where the evolution of each distribution function, giving the probability of the simultaneous distribution of n particles, depends on that of n + 1 particles. Thus we need to close this hierarchy in some way. This is usually done by expanding in the plasma parameter

$$g=\frac{1}{n{\lambda_d}^3}\quad;\quad\lambda_d=\sqrt{\frac{T}{4\pi en}}$$

Which is the inverse number of particles in a Debyesphere. When the plasmaparameter tends to zero only collective interactions remain between the particles. The effect is as if the particles were smeared out in phase space. When we study the equation of the one particle distribution function and include effects of the two particle distribution function (describing pair collisions) as expanded in g we get the equation:

$$\left\{\frac{\partial}{\partial t}+\vec{v}\cdot\frac{\partial}{\partial\vec{r}}+\frac{e}{m}\vec{E}\cdot\frac{\partial}{\partial\vec{v}}+\Omega(\vec{v}\times\hat{e}_{||})\cdot\frac{\partial}{\partial\vec{v}}\right\}f(\mathbf{r},\mathbf{v},\mathrm{t})=\left(\frac{\partial f}{\partial t}\right)_{coll} \tag{2.5}$$

where f is the one particle distribution function and the right hand side approximates close collisions (first order in g). Here various approximations like Boltzmanns or the Fokker-Planck collision terms are used. If we can ignore close collisions completely we have the Vlasov equation:

$$\left\{\frac{\partial}{\partial t}+\vec{\mathrm{v}}\cdot\frac{\partial}{\partial\vec{r}}+\frac{e}{m}\vec{E}\cdot\frac{\partial}{\partial\vec{\mathrm{v}}}+\Omega(\vec{\mathrm{v}}\times\hat{e}_{\|})\cdot\frac{\partial}{\partial\vec{\mathrm{v}}}\right\}f(\mathbf{r},\mathbf{v},\mathrm{t})=0 \tag{2.6}$$

2.2 Kinetic Theory as Generally Used by Plasma Physicists

The kinetic equations 2.5 and 2.6 are the equations usually used by plasma physicists. Equation 2.6 is reversible like (2.4). This means that processes can go back and forth. Equation 2.6 describes only collective motions. An example of this is wave propagation. It is also able to describe temporary damping (in the linearized case) of waves, so called Landau damping, due to resonances between particles and waves. Since the plasma parameter g in typical laboratory plasmas is of the order 10^{-8} collective phenomena usually dominate over phenomena related to close collisions. We mentioned above the Fokker-Planck collision term for close collisions. However, in a random phase situation also turbulent collisions can be described by a Fokker-Planck equation. It can be written:

$$\left(\frac{\partial}{\partial t}+\mathrm{v}\frac{\partial}{\partial x}\right)f(x,\mathrm{v},t)=\frac{\partial}{\partial \mathrm{v}}\left[\beta\mathrm{v}+D^{\mathrm{v}}\frac{\partial}{\partial \mathrm{v}}\right]f(x,\mathrm{v},t) \tag{2.7}$$

The Fokker-Planck equation is fundamental and of interest in many contexts. One aspect is that it, in its original form is Markovian (particles have forgotten previous events) but recently has been generalized to the non-Markovian case [23]. For constant coefficients it has an exact analytical solution [1]. It is interesting to note that already two waves leads to stochasticity of particles, giving quasilinear transport [17]. However, the solution of the Fokker-Planck equation, including friction, leads to a saturation of the mean square deviation of velocity after a time of order $1/\beta$. A solution is shown in Fig. 2.1. For the turbulent case (β and D^{v} depending on intensities of the turbulent waves), the initial linear growth of $<(\Delta \mathrm{v})^2>$ is according to quasilinear theory and would be predicted by the Chirikov results. The saturation follows from Dupree-Weinstock theory [12, 13] which is a strongly nonlinear renormalized theory. Thus in the flat region nonlinearities have introduced correlations. This is analogous to correlations between three wave packets introduced by nonlinearities in the Random Phase approximation [11].

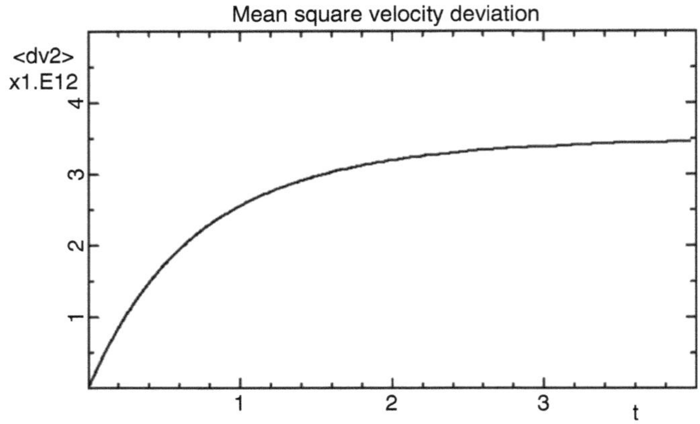

Fig. 2.1 Mean square velocitydeviation $<(\Delta v)^2>$ as a function os time showing intitial quasilinear linear growth and later saturation at t ~1/β

2.3 Gyrokinetic Theory

For low frequency ($\omega << \Omega_c$) we can average the kinetic equation over the gyromotion. This leads to the gyrokinetic equation:

$$\left(\omega - \omega_D\left(v_{||}^2, v_\perp^2\right) - k_{||}v_{||}\right)\left(f^{(1)}{}_{k,\omega} + \frac{q\phi_{k,\omega}}{T}f_0\right)e^{-iL_k}$$
$$= \left[(\omega - \omega_\bullet)\frac{q}{T}(\phi_{k,\omega} - v_{||}\, A_{||})J_0(\xi_k) - i\frac{v_\perp}{k_\perp}(\hat{e}_{||} \times \mathbf{k}) \cdot \mathbf{A_k} J_0{}'\right]f_0 \tag{2.8a}$$

We will return to the derivation of this equation, and its nonlinear extension in Chap. 5.

However, we will mention that the magnetic drift is the sum of gradient B and curvature drifts as:

$$\mathbf{v}_D = \mathbf{v}_{\nabla B} + \mathbf{v}_\kappa \tag{2.8b}$$

where

$$\mathbf{v}_{\nabla \mathbf{B}} = \frac{v_\perp{}^2}{2\Omega_c}(\hat{e}_{||} \times \nabla \ln B) \tag{2.8c}$$

$$\mathbf{v}_\kappa = \frac{v_{||}{}^2}{\Omega_c}(\hat{e}_{||} \times \kappa) \tag{2.8d}$$

and

$$\kappa = (\hat{e}_{\|} \cdot \nabla)\hat{e}_{\|} \tag{2.8e}$$

The drifts defined by (2.8c) and (2.8d) are the gradB and curvature drifts respectively and (2.8e) defines the curvature vector. We here use $\hat{e}_{\|} = B/B$ as a space dependent unity vector along the magnetic field. In a slab geometry with fixed magnetic field we instead use the unity vector $\hat{\mathbf{z}}$.

Since $\hat{e}_{\|}$ it is generally used in inhomogeneous systems we will need to solve eigenvalue equations. Then $k_{\|}$ and sometimes even the magnetic drift frequency will become operators. Since then the eigenvalue problem depends on the particular velocity we are considering, the total eigenvalue solution will have to be averaged over velocity space. Thus we have an integral eigenvalue problem. The fact that magnetic curvature is destabilizing on the outside and stabilizing on the inside of a torus will show in a dependence of ω_D on the poloidal angle. The density perturbation from (2.8a) will be obtained by dividing by the first factor and integrating over velocity.

$$\frac{\delta n_i}{n_i} = -\frac{e\phi}{T_i}\left[1 - \frac{1}{n_0}\int_0^\infty \frac{\omega - \omega_{*i}[1 + \eta_i(m_i v^2/2T_i - 3/2]}{\omega - k_{\|}v_{\|} - \omega_{D_i}\left(v_{\|}^2 + v_{\perp}^2/2\right)/v_{th}^2} J_0(\xi)^2 f_0 d^3v\right] \tag{2.9}$$

We here took the electrostatic approximation just for the purpose of illustration. The integral in (2.9) will have resonances corresponding to wave particle resonances. However, as will be discussed later, in the nonlinear regime, nonlinear frequency shifts may detune these resonances.

2.4 Fluid Theory as Obtained by Taking Moments of the Vlasov Equation

An alternative to making the full kinetic calculation is to first derive fluid equations by taking moments of (2.5) or (2.8a) (of course collisions can be added also to [2.8]). Clearly, in general (2.9) contains less information than (2.5). However, if we expand the fluid equations obtained from (2.5) in the low frequency limit the results obtained from (2.5) and (2.8a) the results will be identical. The equations obtained by taking moments of (2.5) are called *fluid equations* and the equations obtained by taking moments of (2.8a) are called *gyrofluid equations.*

Fluid equations really describe a continuum where the local velocities have been averaged over the particle distribution at every point. This leads to the presence of **fluid drifts** that are **not guiding centre drifts** in an inhomogeneous plasma. However, the macroscopic properties like the time derivative of the density are, of course, the same whether we use fluid or gyrofluid equations. Another aspect which is not either a really dividing property is the fact that several authors have added the linear kinetic resonances to gyrofluid equations. These are then called Gyro-Landau Fluid resonances. However, this is just a question of habits of different authors and, of course, there is nothing that prevents us from adding linear kinetic resonances to fluid equations.

2.4.1 *The Maxwell Equations*

Since the ordinary fluid equations are what we will mainly use in this book we will here start by including the Maxwells equations.

$$\nabla \times \mathbf{E} = -\frac{\partial \mathbf{B}}{\partial \mathbf{t}} \tag{2.10a}$$

$$\nabla \times \mathbf{B} = \mu_0 \mathbf{J} + \frac{\partial \mathbf{D}}{\partial \mathbf{t}} \mu_0 \tag{2.10b}$$

$$\nabla \cdot \mathbf{B} = 0 \tag{2.10c}$$

$$\nabla \cdot \mathbf{E} = \frac{\rho}{\varepsilon_0} \tag{2.10d}$$

Here (2.10a) is the induction law and (2.10b) is the ampere law. Here the last term is the displacement current which will generally be neglected here since we consider low frequencies where quasineutrality holds. Equation 2.10c is general and tells us that there are no magnetic charges while (2.10d) will mostly be replaced by the quasineutrality condition.

2.4.2 *The Low Frequency Expansion*

$$\frac{\partial \mathbf{v}}{\partial t} + (\mathbf{v} \cdot \nabla)\mathbf{v} = \frac{q}{m}(\mathbf{E} + \mathbf{v} \times \mathbf{B}) - \frac{1}{mn}(\nabla P + \nabla \cdot \pi) + \vec{g} = 0 \tag{2.11a}$$

$$\omega << \Omega_c \Rightarrow$$

$$\mathbf{v}_\perp = \mathbf{v}_E + \mathbf{v}_p + \mathbf{v}_* + \vec{\mathbf{v}}_a + \vec{\mathbf{v}}_g \tag{211b}$$

$$\mathbf{v}_E = \frac{1}{B}(\mathbf{E} \times \hat{\mathbf{z}}) \tag{2.11c}$$

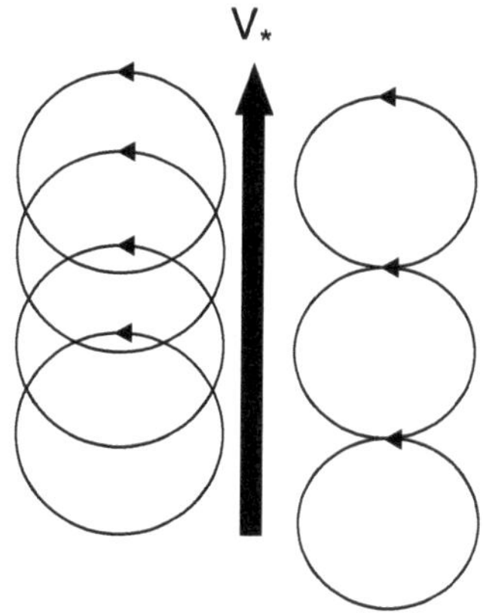

Fig. 2.2 Diamagnetic drift

$$\mathbf{v}_* = \frac{1}{qnB}(\widehat{\mathbf{z}} \times \nabla P) \tag{2.11d}$$

$$\mathbf{v}_p = \frac{1}{\Omega_c}\left(\frac{\partial}{\partial t} + \mathbf{v}\cdot\nabla\right)(\widehat{\mathbf{z}} \times \mathbf{v}) \tag{2.11e}$$

A usual approximation is to substitute the E $\times$ B drift into (2.11e)

This gives:

$$\mathbf{v}_p = \frac{1}{\mathrm{B}\Omega_c}\left(\frac{\partial}{\partial t} + \mathbf{v}_E\cdot\nabla\right)\mathbf{E} \tag{2.11f}$$

$$\vec{v}_g = \frac{\vec{g}\times\widehat{\mathbf{z}}}{\Omega_c} \tag{2.11g}$$

However, this needs to be generalized when we include Finite Larmor Radius (FLR) effects.

Due to the bending of field lines we also have an electromagnetic drift

$$\mathbf{v}_{\delta\mathrm{B}} = \mathrm{v}_{||}\frac{\delta\mathbf{B}_{\perp}}{\mathbf{B}_{||}} \tag{2.11h}$$

Here the diamagnetic drift is a pure fluid drift, i.e. it does not move particles (Fig. 2.2).

Since the diamagnetic drift does not move particles it does not cause a density perturbation i.e. (Fig. 2.2)

$$\nabla\cdot(\mathrm{n}\mathbf{v}_*) = 0 \tag{2.12}$$

Equation 2.12 is the lowest order consequence of the fact that the diamagnetic drift does not move particles. In the momentum equation the stress tensor cancels

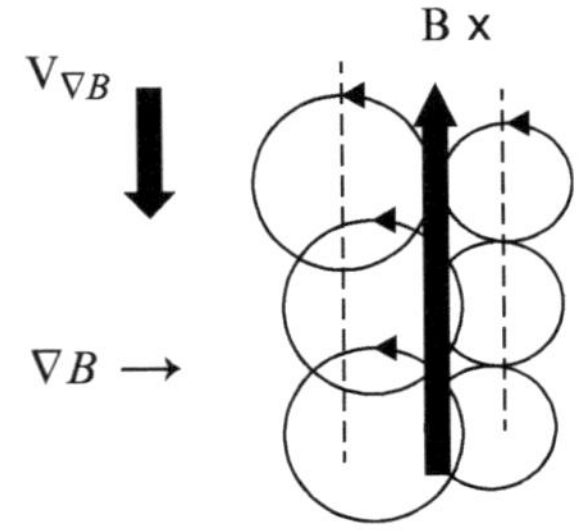

Fig. 2.3 Magnetic drift. The particle drift is compensated by the fact that more particles contribute from the side with weaker magnetic field in such a way that there is no fluid drift

convective diamagnetic effects. Such effects are cancelled also in the energy equation as we will soon see.

The magnetic drift is not a fluid drift because the guiding centre drift is compensated by the fluid effect of having more particles from one side (Fig 2.3)

2.4.3 The Energy Equation

The highest order moment equation that we are going to make use of is the energy equation. It is most commonly written as an equation for the pressure variation as:

$$\frac{3}{2}\left(\frac{\partial}{\partial t} + \mathbf{v_j} \cdot \nabla\right)\mathrm{P_j} + \frac{5}{2}\mathrm{P_j}\nabla \cdot \mathbf{v_j} = -\nabla \cdot \mathbf{q_j} + \sum_{\mathrm{j=i}} \mathrm{Q_{ji}} \tag{2.13}$$

where $\mathbf{q}_j$ is the heat flux and Q_{ij} is the heat transferred from species i to species j by means of collisions. This energy exchange typically contains effects like Ohmic heating and temperature equilibrium terms. It will be neglected in the following. The heat flux q is for the collision dominated case ($\lambda >> l_f$) according to Braginskii:

$$\mathbf{q_j} = 0.71 n_j T_j \mathbf{U}_{||} - \kappa_{||}\nabla_{||}T - \kappa_{\perp}\nabla_{\perp}T + \mathbf{q_{*i}} + \frac{3}{2}\nu_j\frac{n_j T_j}{\Omega_{cj}}(\widehat{e}_{||} \times \mathbf{U}) \tag{2.14}$$

where $\mathbf{U}$ is the relative velocity between species j and i. The thermal conductivities for electrons are given by

$$\kappa_{||e} = 3.16\frac{n_e T_e}{m_e \nu_e} \qquad \kappa_{\perp e} = 4.66\frac{n_e T_e \nu_e}{m_e \Omega_{ce}{}^2}$$

and for ions by

$$\kappa_{||i} = 3.9\frac{n_i T_i}{m_i \nu_i} \qquad \kappa_{\perp i} = 2\frac{n_i T_i \nu_i}{m_i \Omega_{ci}{}^2}$$

and

$$\mathbf{q}_{*j} = \frac{5}{2}\frac{\mathrm{P_j}}{\mathrm{m_j}\Omega_{\mathrm{c_j}}}(\mathbf{e}_{||} x \nabla \mathrm{T_j}) \tag{2.15}$$

If we neglect the full right hand side of (2.14) we obtain the adiabatic equation of state for three dimensional motion, i.e.

$$\frac{d}{dt}\left(Pn^{5/3}\right) = 0 \tag{2.16}$$

which holds for processes that are so rapid that the heat flux does not have time to develop. When div $\mathbf{v} = 0$ which is a rather common situation, the pressure perturbation can be taken as due to convection in a background gradient. This will be further discussed later.

Another usual form of the energy equation is that obtained after subtracting the continuity equation. It may be written as:

$$\frac{3}{2}n_j\left(\frac{\partial}{\partial t} + \mathbf{v_j} \cdot \nabla\right)\mathrm{T_j} + \mathrm{P_j}\nabla \cdot \mathbf{v_j} = -\nabla \cdot \mathbf{q_j} \tag{2.17}$$

Equations 2.13 and 2.17 are fluid equations and the velocities thus contain the diamagnetic drifts. As it turns out these drifts cancel in a way similar to that in the momentum equation but now due to the heat flow terms, i.e.

$$\frac{3}{2}n\mathbf{v}_* \cdot \nabla T - T\mathbf{v}_* \cdot \nabla n = \frac{5}{2}n\mathbf{v}_* \cdot \nabla T \tag{2.18a}$$

$$\frac{3}{2}n\mathbf{v}_* \cdot \nabla T - T\mathbf{v}_* \cdot \nabla n = -\nabla \cdot \mathbf{q}_* \tag{2.18b}$$

Where $\omega_\bullet = \mathrm{k} \bullet \mathrm{v}_\bullet$. We can then write the energy equation in the form:

$$\frac{3}{2}n_j\left(\frac{\partial}{\partial t} + \mathbf{v}_{gc_j} \cdot \nabla\right)\mathrm{T_j} - \mathrm{T_j}\left(\frac{\partial n_j}{\partial t} + \mathbf{v}_{\mathrm{gc_j}} \cdot \nabla n_j\right) = -\nabla \cdot \mathbf{q}_{gcj} \tag{2.19}$$

Where $\mathbf{v}_{\mathrm{gc}}$ here is defined as the guiding centre part of the fluid velocity, i.e. without the magnetic drift and $\mathbf{q}_{\mathrm{gcj}}$ is $\mathbf{q}_{\mathrm{j}}$ as defined in (2.14) but without the diamagnetic heatflow, i.e. (2.15). As we will see later, in a curved magnetic field also (2.15) will contain a guiding centre part. Equation 2.19 shows that the relevant convective velocity in the energy equation is the guiding centre part of the fluid velocity. The term coming from div $\mathbf{v}$ is

$$\frac{\partial n_j}{\partial t} + \mathbf{v}_{\mathrm{gc_j}} \cdot \nabla n_j = -n\nabla \cdot \mathbf{v}_{\mathrm{gcj}} - \nabla \cdot (n\mathbf{v}_{*j})$$

Where the last term is a pure magnetic drift effect. From this follows also that the convective velocity in (2.16) does not contain the diamagnetic drift.

Another useful equation of state may be obtained at low frequencies and small collision rates for electrons. In this case the energy equation is dominated by the div **q** term so that the lowest order equation of state is $\mathbf{q} = 0$ or $\kappa_{||}\nabla_{||}T = 0$. Now $\nabla_{||} = (1/B)(\mathbf{B_0} + \delta\mathbf{B}) \cdot \nabla$ so that, after linearization

$$\mathbf{B_0} \cdot \nabla\delta T + \delta\mathbf{B} \cdot \nabla T_0 = 0 \tag{2.20}$$

If the perpendicular perturbation in **B** is represented by a parallel vector potential we obtain the equation of state:

$$\delta T_j = -\eta_j \frac{\omega_{*j}}{k_{||}} q_j A_{||} \tag{2.21}$$

where $\eta_j = d \ln T_j / d \ln n_j$

Although the above expression for **q** has been derived by assuming domination of collisions along $B(\lambda >> l_f)$ the equation of state (2.21) can also be used to reproduce the electron density response in the limit $\omega << k_{||}v_{||}$ obtained from the Vlasov equation. The reason for this is that it arises as a limiting case that does not depend on the explicit form of $\kappa_{||}$.

With regard to the cancellation of the diamagnetic drifts this effect is very important for vortex modes since typically the perturbed part of $\mathbf{v_*}$ is of the same order as $\mathbf{v_E}$. The application of (2.16) for such modes thus depends strongly on this cancellation and the relevant convective velocity in d/dt is the guiding centre part of the fluid velocity.

2.5 Gyrofluid Theory as Obtained by Taking Moments of the Gyrokinetic Equation

We will now consider equations obtained by taking moments of (2.8a). These are in principle equivalent to fluid equations. Finite Larmor Radius (FLR) effects are included to all orders in gyrofluid equations already at taking the moments while FLR effects in fluid equations have to be obtained by extensive work with convective diamagnetic and stress tensor effects. We refer the reader to Ref [25] in order to see how FLR effects are included in gyrofluid theory. An important difference between gyrofluid and fluid equations is that gyrofluid equations do not contain the pressure term perpendicular to the magnetic field. This simplifies a lot although as mentioned above, taking the moments of the gyrokinetic equation, involving magnetic drifts and Bessel functions is more complicated in itself.

Averaging the magnetic drift (2.8b) over a Maxwellian velocity distribution we get:

$$\mathbf{v}_D = \mathbf{v}_{\nabla B} + \mathbf{v}_\kappa \tag{2.22a}$$

where

$$\mathbf{v}_{\nabla\mathbf{B}} = \frac{\mathrm{T}}{m\,\Omega_c}(\widehat{e}_{||} \times \nabla \ln \mathrm{B}) \tag{2.22b}$$

$$\mathbf{v}_{\kappa} = \frac{\mathrm{T}}{m\,\Omega_c}(\widehat{e}_{||} \times \kappa) \tag{2.22c}$$

The main particle drifts in gyrofluid theory are the ExB drift, $\mathbf{v_E}$, the polarization drift, $\mathbf{v_p}$ and the magnetic drift $\mathbf{v_D}$. Gyrofluid theory is a theory for the motion of guiding centres so diamagnetic or stress tensor drifts are not present. While the perpendicular motion is pretty much given by the drifts just mentioned (the Coriolis drift is added in combination with a toroidal flow) the parallel motion (without flow) is given by (2.23) [25]. This equation is interesting since the parallel motion should be the same for guiding centres and ordinary fluid while fluid equations do not have a convective magnetic drift.

$$\frac{\partial \delta \mathrm{u}_{||}}{\partial \mathrm{t}} + 2\mathbf{v_D} \cdot \nabla \delta \mathrm{u}_{||} = -\widehat{e}_{||} \cdot \nabla(\delta p + en\phi) \tag{2.23}$$

2.6 One Fluid Equations

A characteristic property of the low frequency expansion of the two fluid equations (2.11a–h) is that the dominant guiding centre drift, the $\mathbf{E} \times \mathbf{B}$ drift, is the same for electrons and ions. Thus in some sense we expect the plasma to move as one fluid. Now we know that this can only be an approximation since the drift velocities due to pressure gradients are different for electrons and ions. However, for the strong, global, Magnetohydrodynamic instabilities, the instability is much faster than the drift frequencies introduced by the density and temperature gradients. In this limit it can be useful to introduce one fluid equations. These are derived by adding or subtracting the equations for electrons and ions after multiplication by the respective masses. If course, this is a formal procedure that can be used to introduce also the individual drift motions of ions and electrons. Then, however, the equations are no longer one fluid equations. The basic one fluid equations are:

$$\rho \frac{d\mathbf{v}}{dt} = \mathbf{J} \times \mathbf{B} - \nabla \mathrm{P} \tag{2.24a}$$

$$\mathbf{E} + \mathbf{v} \times \mathbf{B} = \eta\,\mathbf{J} \tag{2.24b}$$

$$\frac{d}{dt}(Pn^{\gamma}) = 0 \tag{2.24c}$$

Here we used the convective derivative $d/dt = \partial/\partial t + v \bullet \text{grad}$, ρ is the mass density, η is the conductivity and γ is the adiabaticity index usually taken as 5/3. Equation 2.24a is the equation of motion, (2.24b) is usually called Ohms law and (2.24c) is the equation of state.

Here (2.24a) retains both ion and electron inertia although electron inertia can almost always be ignored. Ion inertia corresponds to including the ion polarization drift in the two fluid equations. The one fluid equations have been used extensively in order to determine MHD stability of various magnetic configurations. In particular an energy principle method was introduced which was used for pioneering work in the beginning of plasma fusion research.

In the present book we will consider both the global MHD instabilities and microinstabilities important for transport. Since two fluid, or kinetic descriptions, will be needed for microinstabilities, it will thus be more convenient to use a two fluid approach in order to obtain a unified description.

2.7 Finite Larmor Radius Effects in a Fluid Description

Up to now we have neglected diamagnetic contributions to the polarization drift and the stress tensor drift. As it turns out these are related to finite Larmor radius (FLR) effects. We shall show here how the lowest order FLR effects can be obtained by a systematic inclusion of these terms.

We will initially for simplicity neglect temperature gradients and temperature perturbations. This leads to the relation

$$\nabla \cdot \mathbf{v}_* = \frac{T}{qB} \nabla \cdot (\widehat{\mathbf{z}} \times \nabla n/n) = 0 \tag{2.25}$$

Since also $\nabla \cdot \mathbf{v}_0 = 0$ we can to leading order use the incompressibility condition $\nabla \cdot \mathbf{v}_0 = 0$ when substituting drifts into $\mathbf{v_p}$ and $\mathbf{v}_\pi$. We will also assume large mode numbers, i.e. $k >> \kappa = d \ln\ n_0/dx$ and $d\kappa/dx = 0$.

From the stress tensor as given by Braginslii we can obtain effects of viscosity related to friction between particles and collisionless gyroviscosity, which is a pure FLR effect.

The relevant gyroviscous components are:

$$\pi_{xy} = \pi_{yx} = \frac{nT}{2\Omega_c}\left(\frac{\partial v_x}{\partial x} - \frac{\partial v_y}{\partial y}\right) + \frac{1}{4\Omega_c}\left(\frac{\partial q_x}{\partial x} - \frac{\partial q_y}{\partial y}\right) \tag{2.26a}$$

$$\pi_{yy} = -\pi_{xx} = \frac{nT}{2\Omega_c}\left(\frac{\partial v_y}{\partial x} + \frac{\partial v_x}{\partial y}\right) + \frac{1}{4\Omega_c}\left(\frac{\partial q_x}{\partial y} + \frac{\partial q_y}{\partial x}\right) \tag{2.26b}$$

Here **q** is determined by the fluid truncation and will include higher order FLR effects. We note, however, that the part of $\mathbf{q}_*$ corresponding to a flux of perpendicular energy is (compare Eq. (6.30))

$$\mathbf{q}_*{}^{\perp} = 2\frac{\mathrm{P_j}}{\mathrm{m_j}\Omega_{\mathrm{c_j}}}(\mathbf{e}_{\|}x\nabla \mathrm{T}_{\perp})$$

We will start by including only density background gradient

$$\nabla\cdot(\pi)_x = \frac{\partial \pi_{\mathrm{xx}}}{\partial \mathrm{x}} + \frac{\partial \pi_{\mathrm{xy}}}{\partial \mathrm{y}} = -\frac{nT}{2\Omega_c}\Delta \mathrm{v_y} - \frac{T}{2\Omega_c}\left(\frac{\partial \mathrm{v_y}}{\partial \mathrm{x}} + \frac{\partial \mathrm{v_x}}{\partial \mathrm{y}}\right)\frac{dn}{dx}$$

$$\nabla\cdot(\pi)_y = \frac{\partial \pi_{\mathrm{yx}}}{\partial \mathrm{x}} + \frac{\partial \pi_{\mathrm{yy}}}{\partial \mathrm{y}} = \frac{nT}{2\Omega_c}\Delta \mathrm{v_x} + \frac{T}{2\Omega_c}\left(\frac{\partial \mathrm{v_x}}{\partial \mathrm{x}} - \frac{\partial \mathrm{v_y}}{\partial \mathrm{y}}\right)\frac{dn}{dx}$$

These equations can be written in a more compact form as:

$$\nabla\cdot(\pi) = \frac{nT}{2\Omega_c}\left[\widehat{\mathbf{z}}\times\Delta_{\perp}\mathbf{v} + \kappa(\nabla \mathrm{v_y} - \widehat{\mathbf{z}}\times \mathrm{v_x})\right]$$

We now obtain:

$$\mathbf{v}_\pi = \frac{1}{enB}\frac{nT}{2\Omega_c}\widehat{\mathbf{z}}\times\nabla\cdot\pi = -\frac{1}{4}\rho^2\Delta_{\perp}\mathbf{v} + \frac{1}{4}\rho^2\kappa(\widehat{\mathbf{z}}\times \mathrm{v_y} + \nabla \mathrm{v_x})$$

Here ρ is the gyroradius of a general species. Since we are usually interested in substituting our drifts into the equation **div j** = 0 we need to calculate expressions of the form **div**(nv). We then find, including only linear terms in **κ**.

$$\begin{aligned}\nabla\cdot(n\mathbf{v}_\pi) &= \mathbf{v}_\pi\cdot\nabla n_0 + n_0\nabla\cdot\mathbf{v}_\pi \\ &= -\frac{1}{4}\rho^2\nabla n_0\cdot\Delta_{\perp}\mathbf{v} - \frac{1}{4}\rho^2 n_0\Delta_{\perp}\nabla\cdot\mathbf{v} + \frac{1}{4}\rho^2\kappa n_0\Delta_{\perp}\mathrm{v_x}\end{aligned}$$

Now assuming **div v** = 0 We obtain:

$$\nabla\cdot(n\mathbf{v}_\pi) = -\frac{1}{2}\rho^2\nabla n_0\cdot\Delta\mathbf{v} \tag{2.27}$$

The polarization drift can be written in the form:

$$\mathbf{v}_p = \frac{1}{\Omega_c}\left(\frac{\partial}{\partial t} + \mathbf{v}\cdot\nabla\right)(\widehat{\mathbf{z}}\times\mathbf{v})$$

We start by observing that due to our large mode number approximation only perturbed drifts will enter in the last **v**. Then in the linear approximation the **v** term in the convective derivative can only be a background **v**. The only background **v** that we are interested in here is the diamagnetic drift. We will then start by considering the contribution from this term to **div**($n\mathbf{v}_p$). It is:

$$\frac{n}{\Omega_c}(\mathbf{v}_* \cdot \nabla)\nabla \cdot (\widehat{\mathbf{z}} \times \mathbf{v}) = \frac{n}{\Omega_c}(\mathbf{v}_* \cdot \nabla)\left(\frac{\partial v_x}{\partial y} - \frac{\partial v_y}{\partial x}\right)$$
$$= -\frac{1}{2}\kappa n \rho^2 \frac{\partial}{\partial y}\left(\frac{\partial v_x}{\partial y} - \frac{\partial v_y}{\partial x}\right) \qquad (2.28a)$$

Now adding (2.27) and (2.28a) we find (with $\nabla n_0 = -\kappa n_0 \widehat{\mathbf{x}}$):

$$\nabla \cdot (n\mathbf{v}_\pi) + \nabla \cdot \left(\frac{n}{\Omega_c}(\mathbf{v}_* \cdot \nabla)\nabla \cdot (\widehat{\mathbf{z}} \times \mathbf{v})\right)$$
$$= \frac{1}{2}\rho^2 \kappa n \Delta v_x - \frac{1}{2}\kappa n \rho^2 \frac{\partial}{\partial y}\left(\frac{\partial v_x}{\partial y} - \frac{\partial v_y}{\partial x}\right) = \frac{1}{2}\rho^2 \kappa n \frac{\partial}{\partial x}\nabla \cdot \mathbf{v} = 0 \qquad (2.28b)$$

We thus find that *convective diamagnetic contributions to* **div**($n\mathrm{v}_p$) *are exactly cancelled by the stress tensor contribution* **div**($n\mathbf{v}_\pi$).

This result can easily be understood from a physical point of view since the diamagnetic drift is not a particle drift and cannot transfer information by convection. *We now have the general result*:

$$\nabla \cdot \left[n(\mathbf{v}_\mathbf{p} + \mathbf{v}_\pi)\right] = \nabla \cdot \left[\frac{n}{\Omega_c}\frac{\partial}{\partial t}(\widehat{\mathbf{z}} \times \mathbf{v})\right] \qquad (2.29)$$

It is also interesting to compare (2.29) with (2.12). In order to obtain a result corresponding to (2.12) for the gyroviscous part of the stress tensor drift $\mathbf{v}_\pi$ it is necessary to add the convective diamagnetic parts of the polarization drift which are of the same order in $k^2\rho^2$. We may thus consider (2.29) to express the same kind of physics as (2.12) but for drifts that are first order in the FLR parameter $k^2\rho^2$. Since (2.12) is no longer true in the presence of curvature (compare 6.23) the same is expected for (2.29).

The leading order linear contributions to (2.29) are now:

$$\frac{n}{\Omega_{ci}}\frac{\partial}{\partial t}\nabla \cdot (\widehat{\mathbf{z}} \times \mathbf{v}_E) = -\frac{1}{2}n\rho_i^2\frac{\partial}{\partial t}\Delta\frac{e\phi}{T_i}$$

$$\frac{n}{\Omega_{ci}}\frac{\partial}{\partial t}\nabla \cdot (\widehat{\mathbf{z}} \times \mathbf{v}_{*i}) = -\frac{1}{2}n\rho_i^2\frac{\partial}{\partial t}\Delta\delta n$$

Here only the perturbation in density contributes to the last term. We now have to specialize further to a particular density response. For flute modes, which are of particular interest in this context, the simplest leading order density perturbation is the E $\times$ B convective, i.e.

$$\frac{\delta n}{n} = \frac{\omega_{*e}}{\omega}\frac{e\phi}{T_e} \tag{2.30}$$

We then obtain in (ω,k) space:

$$\nabla \cdot \left[n(\mathbf{v_p} + \mathbf{v}_\pi)\right] \approx -\frac{i}{2}nk^2\rho_i{}^2(\tau\omega + \omega_{*e})\frac{e\phi}{T_e} = -ink^2\rho^2(\omega - \omega_{*i})\frac{e\phi}{T_e} \tag{2.31}$$

The result (2.31) is in agreement with kinetic theory (compare Chap. 4). The FLR effect enters as a convective contribution to the polarization drift but is in fact due to the time variation of the perturbed diamagnetic drift.

2.7.1 *Effects of Temperature Gradients*

The main source of modification in the presence of temperature gradients is a compressibility of $\mathbf{v}_*$. Thus (2.24) is changed into:

$$\nabla \cdot \mathbf{v}_* = \frac{en}{qB}\nabla \cdot \left[\frac{1}{n}(\hat{\mathbf{z}} \times \nabla \mathrm{T})\right] \tag{2.32}$$

Since (2.25) contains n only in the combination P = nT, (2.26) remains unchanged if we change the definition of κ into $\kappa_\mathrm{p} = -(1/\mathrm{P}_0)\mathrm{dP}_0/\mathrm{dx}$. We then have:

$$\begin{aligned}\nabla\cdot(n\mathbf{v}_\pi) &= -\frac{1}{4}\rho^2\nabla n_0 \cdot \Delta_\perp \mathbf{v} - \frac{1}{4}\rho^2 n_0 \Delta_\perp \nabla \cdot \mathbf{v} + \frac{1}{4}\rho^2\kappa n_0 \Delta_\perp \mathrm{v_x} \\ &-\frac{1}{4}\rho^2 n_0\frac{\nabla T}{T}\cdot \Delta_\perp \mathbf{v} - \frac{1}{4m\Omega_c{}^2}\Delta_\perp \nabla \cdot \mathbf{q}_*{}^\perp\end{aligned} \tag{2.33}$$

where the last term is due to the **q** parts of (2.25a,b). As it turns out it cancels the **div v.** term. Thus (2.27) is changed into:

$$\nabla \cdot (n\mathbf{v}_\pi) = -\frac{1}{2}\rho^2\frac{1}{T}\nabla P_0 \cdot \Delta \mathbf{v} \tag{2.34}$$

Since, in the presence of temperature gradients, **v.** in the convective derivative of the polarization drift contains the full pressure gradient we now find that (2.28b) is unchanged (with our new definition of κ) and so is the conclusion in italics following it and (2.29). Since background pressure gradients in a natural way lead to convective pressure perturbations we now must write:

$$\frac{n}{\Omega_{ci}}\frac{\partial}{\partial t}\nabla \cdot (\hat{\mathbf{z}} \times \mathbf{v}_*) = -\frac{1}{2}\frac{\rho_i{}^2}{T}\frac{\partial}{\partial t}\Delta\delta P \tag{2.35}$$

For the convective pressure perturbation we have:

$$\frac{\delta P_i}{n} = -\frac{\omega_{*iT}}{\omega}\frac{e\phi}{T_i} \tag{2.36}$$

where ω_{*iT} is the diamagnetic drift frequency of ions due to the full background pressure gradient. Accordingly, (2.31) becomes:

$$\nabla \cdot \left[n(\mathbf{v_p} + \mathbf{v}_\pi)\right] = -ink^2\rho^2(\omega - \omega_{*iT})\frac{e\phi}{T_e} \tag{2.37}$$

References

1. S. Chandrasekhar, *Stochastic Problems in Physics and Astronomy*, Rev. Modern Physics **15**, 1 (1943).
2. H. Alfve'n, Cosmical Electrodynamics, Oxford University Press, 1950.
3. S. Chapman and T. G. Cowling, The Mathematical Theory of Non-uniform Gases, Cambridge, London 1958.
4. W.B. Thompson, An Introduction to Plasma Physics, Pergamon Press, Oxford 1964.
5. S.I. Braghinskii, in Reviews of Plasma Physics (e.d. M A Leontovich) Consultants Bureau, New York, Vol. 1, 205 (1965).
6. B.B. Kadomtsev, Plasma Turbulence, Academic Press, New York 1965.
7. A.B. Mikhailovskii and L I Rudakov, Soviet Physics JETP 17, 621 (1963).
8. K.V. Roberts and J.B. Taylor, Phys Rev Lett **8**, 197 (1962)
9. B Lehnert, Dynamics of Charged Particles, North-Holland, Amsterdam 1964.
10. Yu.L. Klimontovich, *Statistical Theory of Nonequilibrium Processes in Plasma*, Pergamon Press, Oxford 1967.
11. R.Z. Sagdeev and A.A. Galeev, *Nonlinear Plasma Theory*, Benjamin, New York 1969.
12. T.H. Dupree, Phys. Fluids 9, 1773 (1966).
13. J. Weinstock, Phys. Fluids 12, 1045 (1969).
14. F.L. Hinton and C W Horton, Phys Fluids **14**, 116 (1971)
15. S.T. Tsai, F W Perkins and TH Stix, Phys Fluids 13, 2108 (1970).
16. N.A. Krall and A W Trivelpiece, Principles of Plasma Physics, McGraw-Hill 1973.
17. B.V. Chirikov, Phys. Rep. **52**, 263 (1979).
18. W. Horton in Handbook of Plasma Physics Vol II ed:s M.N. Rosenbluth and R.Z. Zagdeev (Elsevier Science Publishers, 1984) PP383–449).
19. J.P. Mondt and J. Weiland, Phys. Fluids B**3**, 3248 (1991); Phys. Plasmas 1, 1096 (1994).
20. W.M. Manheimer and C.N. Lashmore Davies, MHD and Microinstabilities in Confined Plasma, (ed. E.W. Laing) Adam Hilger, Bristol 1989.
21. A.G. Sitenko and V. Malnev, *Plasma Physics Theory*, Chapman and Hall, London 1995.
22. R.J. Goldston and P.H. Rutherford, An introduction to Plasma Physics, Adam Hilger, Bristol, 1995.
23. A. Zagorodny and J. Weiland, Physics of Plasmas **6**, 2359 (1999).
24. J. Weiland, Collective Modes in Inhomogeneous Plasma, Kinetic and Advanced Fluid Theory, IoP, Bristol 2000.
25. R.E. Waltz, R.R. Dominguez and G.W. Hammett Phys. Fluids *B4*, 3138 (1992).

Chapter 3
Fluid Description for Low Frequency Perturbations in an Inhomogeneous Plasma

3.1 Introduction

We will now start to apply our fluid equations discussed in Chap. 2 to some fundamental modes in inhomogeneous plasmas. The literature in this field is extensive [1–49]. We will here start by studying the effects of the inhomogeneities themselves, without complicated geometry. We will also usually simplify our description so as to disregard temperature perturbations and background gradients. Such effects are very important but lead to considerably more complicated descriptions and will be considered in Chap. 6.

The main reason for our interest in these modes is their potential importance for anomalous transport and also for more macroscopic convective instabilities as, e.g. the kink instability. Since we are here going to avoid too strong effects of geometry and boundaries we will restrict consideration to the WKB case, i.e. $k_\perp >> \text{grad}(\ln\ n)$ corresponding to large mode numbers in a torus. These modes also have $k_\| << k_\perp$ and if toroidal effects are included they require the solution of an eigenvalue problem along the magnetic field. The effects of this eigenvalue problem will here only be hinted.

Our basic geometry will be that of a plasma slab with the density gradient in the negative x direction and the magnetic field in the positive z direction (Fig. 3.1). In a toroidal machine x corresponds to the radial coordinate, y to the poloidal coordinate and z to the toroidal coordinate. A local mode will have an extent in the radial direction which is much smaller than the typical scale of background variation. The most rapid variation, however, often takes place in the poloidal, y direction and when $k_y >> k_x$ the equations can be conveniently simplified by neglecting k_x as will sometimes be done in the following. This also has the advantage that we avoid the radial eigenvalue problem. Details of eigenvalue problems will be postponed to Chap. 6 (Fig. 3.2).

J. Weiland, *Stability and Transport in Magnetic Confinement Systems*,
Springer Series on Atomic, Optical, and Plasma Physics 71,
DOI 10.1007/978-1-4614-3743-7_3, © Springer Science+Business Media New York 2012

Fig. 3.1 Slab picture of a magnetized plasma with a density gradient

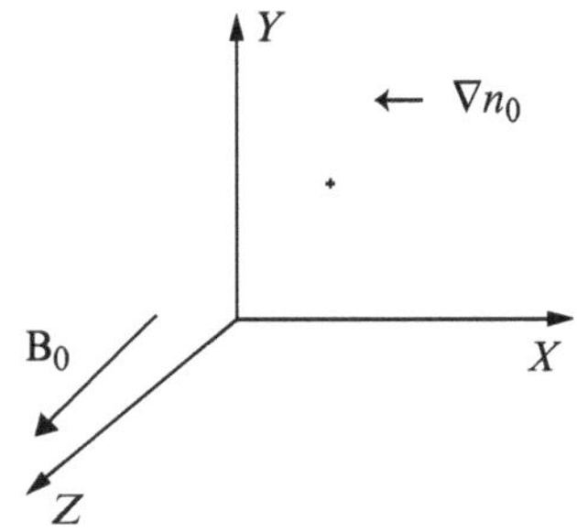

Fig. 3.2 A perturbation following a field line in a torus

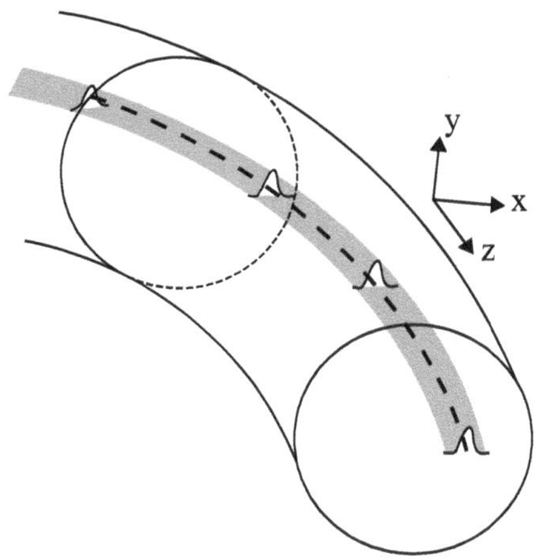

As mentioned in Chap. 2, electron motion along the magnetic field lines has a stabilizing influence on the modes we consider. For small $k_{||}$ the electron motion along the field lines is less efficient for cancelling space charge. This is the reason for our interest in modes with, small $k_{||}$, i.e. we assume:

$$k_{||} << k_{\perp}$$

In this case the main variation of the mode is in the perpendicular plane. The parallel electron motion is quite different for different modes that we will consider in the following. We may here separate two classes. The first class is that of drift waves for which $E_{||} \neq 0$. The second class is the Magnetohydrodynamic (MHD) type modes for which $E_{||} \approx 0$. In the first case the electrons are essentially free to cancel space charge by moving along the magnetic field while in the second case the parallel electron motion is strongly impeded either by a very small $k_{||}$ or by electromagnetic induction. As will be shown in Exercise 8 also the effects of magnetic induction on $E_{||}$ increase for small $k_{||}$ but the direction of propagation (sign of ω) also strongly influences $E_{||}$ which has a maximum close to the electron diamagnetic drift frequency.

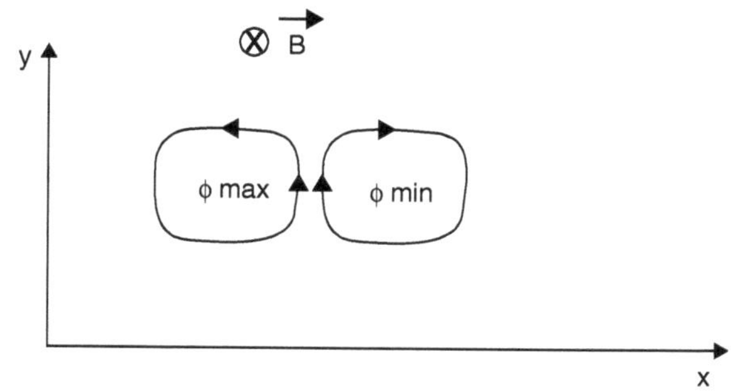

Fig. 3.3 Convective cells

As shown in Chap. 2 a vorticity $\Omega = \nabla \times \vec{v}_E = (1/B_0)\Delta_\perp$ is associated with the perpendicular motion of all these modes. This means that the fluid motion forms rotating whirls. For periodic variation in x and y the velocity typically has a structure as shown in Fig. 3.3 where we have shown one wavelength in the y direction.

The figure shows the characteristic "smoke ring" structure caused by the opposite senses of rotation of the **E** × **B** drift around potential minima and maxima. The actual fluid velocity is that shown in the figure while the structure as such moves with the phase velocity of the wave. It is rather obvious from this picture that vortex modes are strong potential candidates for causing anomalous transport, i.e. the fluid motion (convection) tends to mix regions of higher and lower density. As is intuitively clear, however, if the perturbation is purely harmonic in time and space also the fluid motion will be completely harmonic and no net transport takes place. When there is a net damping or growth, however, this coherent picture is modified and a transport takes place. This will be shown in the end of this chapter as quasilinear diffusion. Of particular interest in connection with convection is the convective cell mode. It has zero real part of the eigenfrequency and thus corresponds to a stationary convection in Fig. 3.3. In this situation a very small irreversible effect in terms of linear damping or growth or spatial "phase mixing" is enough to cause a substantial transport.

3.2 Elementary Picture of Drift Waves

Drift waves are basically electrostatic modes introduced by inhomogeneities in density and as we will show in Chap. 6 in temperature. However, electromagnetic effects on drift waves are often needed and introducing electromagnetic effects will make it possible to make the transition between drift type and MHD type modes. A characteristic feature of drift waves is that their parallel phase velocity is between the ion and electron thermal velocities:

$$v_{thi} \le \frac{\omega}{k_{\|}} << v_{the} \tag{3.1}$$

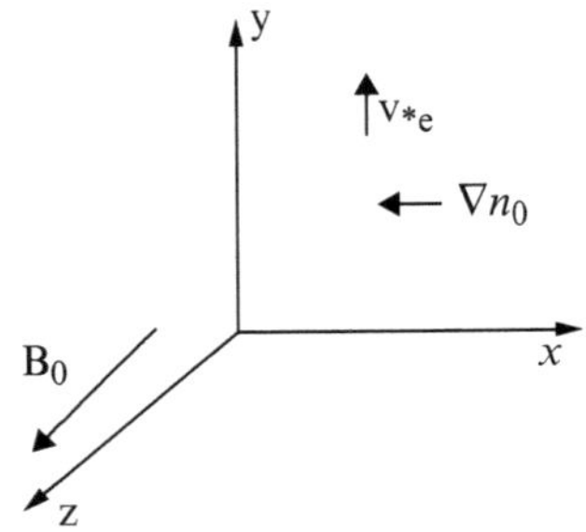

Fig. 3.4 Elementary drift wave geometry

We now specify the background density gradient to be in the negative x direction while the background magnetic field is in the positive z direction.

The zero order diamagnetic drift $\mathbf{v}_{\bullet \mathbf{e}}$ of the electrons due to the background density gradient will then be in the positive y direction and takes the value:

$$\mathbf{v}_{\bullet} = \frac{\kappa \mathrm{T_e}}{\mathrm{eB_0}}$$

where $\kappa = -(1/\mathrm{n_0})\mathrm{dn_0}/\mathrm{dx}$.

In the analysis of low frequency waves, the magnitude of $\mathrm{k}_{\|}$ is very significant. We may write the parallel equation of motion of electrons as

$$m_e \left[\frac{\partial \mathrm{v}_{\|\mathrm{e}}}{\partial \mathrm{t}} + (\mathbf{v_e} \cdot \nabla)\mathrm{v}_{\|\mathrm{e}} \right] = e\frac{\partial \phi}{\partial z} - \frac{1}{n_e}\frac{\partial \mathrm{p_e}}{\partial \mathrm{z}} \tag{3.2}$$

The left hand side of (3.2) is due to electron inertia. It enters only under extreme conditions such as e.g. for the collisionless skindepth or collisionless tearing modes. For drift waves it is neglected due to the electron part of (3.1). For a slow process we can also use an isothermal equation of state. Then (3.2) leads to:

$$e\frac{\partial \phi}{\partial z} = \frac{\mathrm{T_e}}{n_e}\frac{\partial \mathrm{n_e}}{\partial z}$$

Which can be integrated to:

$$\frac{\mathrm{n_e}}{\mathrm{n_0}} = e^{e\phi/\mathrm{T_e}} \tag{3.3a}$$

Equation 3.3a is the Boltzmann distribution which is usually a good approximation if $\mathrm{k}_{\|}$ is not too small. Writing $\mathrm{n_e} = \mathrm{n_0} + \delta \mathrm{n_e}$ and expanding the exponential for $\mathrm{e}\varphi/\mathrm{T_e}$ we find (Figs. 3.4, 3.5):

$$\frac{\delta \mathrm{n_e}}{\mathrm{n_0}} = \frac{e\phi}{\mathrm{T_e}} \tag{3.3b}$$

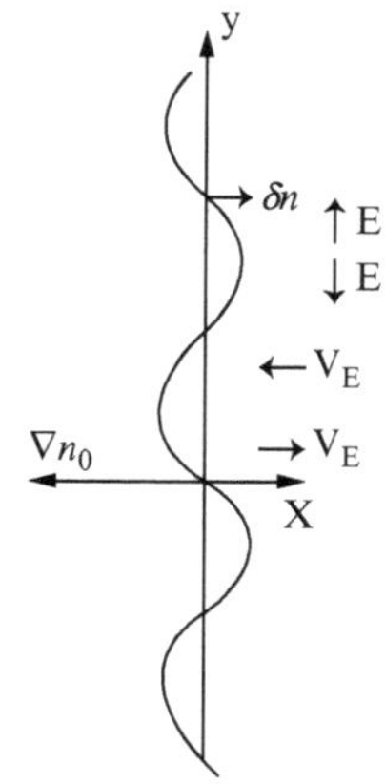

Fig. 3.5 Mechanism of drift wave propagation

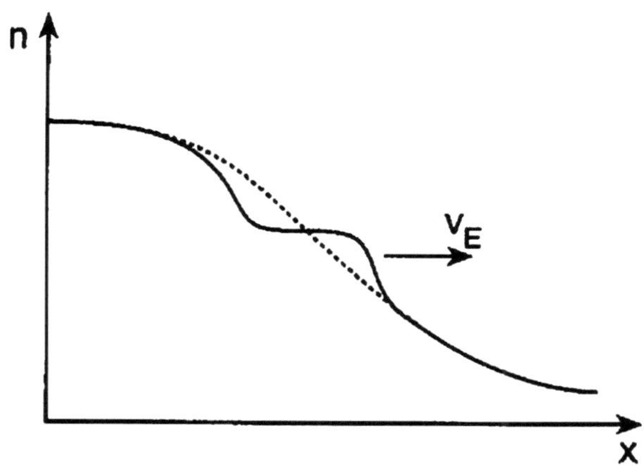

Fig. 3.6 Convective density perturbation

This result is in agreement with our previous estimate for the validity of quasineutrality. This field will cause an $\mathbf{E} \times \mathbf{B}$ drift v_{Ex} in the x direction as shown in Fig. 3.6. This drift will, due to the background density gradient cause a change of density in such a way that the perturbation moves in the positive y direction. Now ignoring magnetic curvature, $\nabla \cdot (n\mathbf{v}_*) = 0$, and $\nabla \cdot (\mathbf{v}_E) = 0$, we obtain from the linearized continuity equation for ions by including only $\mathbf{v_E}$,

$$\frac{\partial n_i}{\partial t} + v_{Ex} \frac{dn_0}{dx} = 0 \tag{3.4}$$

In (3.4) we dropped the parallel motion of the ions. This is permitted for small enough $k_{\|}$ since then the first inequality in (3.1) becomes strong. Equation 3.4 describes the density variation due to convection mentioned above. It corresponds to an incompressible motion Introducing

$$\mathbf{v}_{Ex} = -\frac{1}{B_0}\frac{\partial \phi}{\partial y}$$

We now obtain from (3.4)

$$\frac{1}{n_0}\frac{\partial n_i}{\partial t} + \frac{1}{B_0}\frac{\partial \phi}{\partial y}\frac{1}{n_0}\frac{dn_0}{dx} = 0$$

We now use quasineutrality

$$\frac{\delta n_i}{n_i} = \frac{\delta n_e}{n_e} \tag{3.5}$$

Then using also (3.3b) we arrive at

$$\frac{\partial \phi}{\partial t} + v_{\bullet e}\frac{\partial \phi}{\partial y} = 0 \tag{3.6a}$$

We now assume a perturbations varying sinusoidally in time and along y. We then obtain the simpliest possible dispersion relation for drift waves:

$$\omega = \omega_{\bullet e} \tag{3.6b}$$

where we introduced the electron diamagnetic drift frequency $\omega_{\bullet e} = k_y v_{\bullet e}$.

3.2.1 Effects of Finite Ion Inertia

We are now interested in extending the result of the previous section. As it turns out, effects of ion inertia, which cause the drift motion of electrons and ions to be different, are also associated with compressibility. First we note that the Boltzmann distribution for the electron density (3.3b) is also obtained for ion acoustic waves propagating along B_0 and corresponds to an expansion of the kinetic integral for the density perturbation in the upper limit of (3.1).

If we assume the ion temperature to be very small, so that the region (3.1) usually considered for drift waves is wide, we may drop the ion pressure term and obtain:

$$v_{\|i} = \frac{k_{\|}}{\omega}\frac{e\phi}{m_i} \tag{3.7}$$

Including now also the ion polarisation drift for the perpendicular motion and still assuming $k_x = 0$ we have

$$\mathbf{v}_{\perp i} = ik_y \frac{\phi}{B_0}\widehat{x} - k_y \frac{i}{B_0}\frac{\partial}{\partial t}\frac{\phi}{\Omega_{ci}}\widehat{y} + \mathbf{v}_{\bullet i} \tag{3.8}$$

Introducing now (3.8) into the ion continuity equation assuming $k_y >> \kappa$ using (2.12) and $\mathbf{v}_E = 0$ we find

$$\frac{\delta n_i}{n_0} = \left(\frac{\omega_{\bullet e}}{\omega} - \frac{k_y{}^2 T_e}{m_i \Omega_{ci}{}^2} + \frac{T_i}{m_i}\frac{k_{\|}{}^2}{\omega^2}\right)\frac{e\phi}{T_e} \tag{3.9}$$

Combining now (3.9) with (3.3), using the continuity equation we obtain the dispersion relation:

$$\omega^2\left(1 + k_y{}^2 \rho_s{}^2\right) - \omega\omega_{\bullet e} - k_{\|}{}^2 c_s{}^2 = 0 \tag{3.10}$$

where we introduced

$$\rho_s = \frac{c_s}{\Omega_{ci}} \tag{3.11}$$

and $c_s = T_e/m_i$. The term $k_y \rho_s$ originates from the ion polarisation drift and represents the influence of ion inertia. ρ_s is the ion Larmor radius at the electron temperature. The dispersion relation (3.10) is represented in Fig. 3.7.

Thus we have seen that the polarization drift enters in the same way as an FLR effect. Because of this it is convenient to rewrite the expression (2.11f) for the polarization drift in the electrostatic case as:

$$\mathbf{v}_p = -\rho_s{}^2\left(\frac{\partial}{\partial t} + \mathbf{v}\cdot\nabla\right)\nabla\frac{e\phi}{T_e} \tag{3.12}$$

From Fig. 3.7 we realize that for large $k_{\|}$ the drift wave turns into the ion acoustic wave. Clarly ion parallel motion may be neglected when $k_{\|} c_s << \omega_{\bullet e}$ if $(k_y \rho_s)^2 << 1$.

For comparison we note that for typical JET parameters we have $c_s \approx 10^6$ m/s and $v_{\bullet e} \approx 10^3$ m/s, i.e. we have to require $k_{\|} << k_y 10^{-3}$ in order to drop the parallel ion motion. For the Boltzmann distribution of electrons to be valid we require $k_{\|} v_{the} >> k_y v_{\bullet e}$ which means $k_{\|} >> 0.25\ k_y\ 10^{-4}$ for JET. The interest in such small $k_{\|}$ is mainly due to the fact that instability is likely to occur in this region.

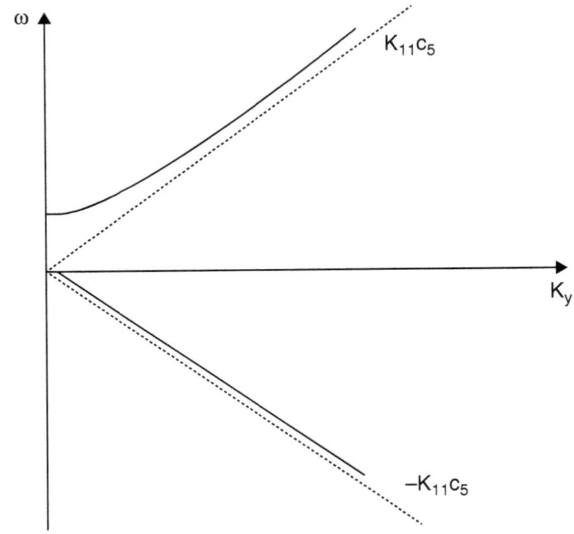

Fig. 3.7 Two dimensional dispersion diagram for drift waves

3.2.2 *Drift Instability*

As long as the electrons are free to move along B_0 to cancel space charge, the Boltzmann relation (3.3) is fulfilled and the drift wave is stable. There are, however, several effects that may limit the mobility of the electrons so as to modify (3.3). These effects are generally more important for small $k_{\|}$ and may be, e.g. electron ion collisions, Landau damping, electron inertia or inductance.

If the electrons are not able to move completely freely there will appear a phase shift, corresponding to a time lag between density and potential in (3.3). We then modify (3.3b) as

$$\frac{\delta n_e}{n_0} = \frac{e\phi}{T_e}(1 - i\delta) \tag{3.13}$$

By replacing (3.3b) with (3.13) in the derivation of (3.6) we obtain the result:

$$\omega = \frac{\omega_{\bullet e}}{(1 - i\delta)} \approx \omega_{\bullet e}(1 + i\delta) \tag{3.14}$$

if we assume $\delta << 1$. We note that due to the time variation $e^{-i\omega t}$, $\delta > 0$ means that the potential lags behind the density. This situation corresponds to an instability.

3.2.3 Excitation by Electron-Ion Collisions

As an example we will now consider the collisional drift instability. We assume the ordering

$$\omega << \nu_{ei} << \Omega_{ci} \tag{3.15}$$

where ν_{ei} is the electron-ion collision frequency. In the regime (3.15) we may for small $k_{\|}$ include the effect of electron ion collisions on the electron parallel motion but continue to drop electron inertia. Dropping the parallel ion motion we then find:

$$v_{\|e} \approx i \frac{k_{\|} T_e}{\nu_{ei} m_e} \left(\frac{e\phi}{T_e} - \frac{\delta n_e}{n_0} \right) \tag{3.16}$$

Taking the limit $k_x = 0$ we have:

$$\mathbf{v}_{\perp \mathbf{e}} = -ik_y \frac{\phi}{B} \widehat{\mathbf{x}} + \mathbf{v}_{\bullet \mathbf{e}}$$

We then get from the electron continuity equation:

$$-i\omega \frac{\delta n_e}{n_0} + ik_y \frac{\kappa T_e}{eB_0} \frac{e\varphi}{T_e} - \frac{k_{\|}{}^2 T_e}{\nu_{ei} m_e} \left(\frac{e\varphi}{T_e} - \frac{\delta n_e}{n_0} \right) = 0$$

Which reduces to:

$$\frac{\delta n_e}{n_0} = \frac{e\phi}{T_e} \frac{\omega_{\bullet e} + ik_{\|}{}^2 D_{\|}}{\omega + ik_{\|}{}^2 D_{\|}} \tag{3.17}$$

where

$$D_{\|} = \frac{T_e}{m_e \nu_{ei}} \tag{3.18}$$

is the parallel diffusion coefficient.

For the orderings already introduced it is reasonable to assume that $k_{\|}{}^2 D_{\|} >> \omega$ Thus expanding (3.17) we obtain:

$$\frac{\delta n_e}{n_0} = \frac{e\phi}{T_e} \left[1 - i \frac{m_e \nu_{ei}}{k_{\|}{}^2 T_e} (\omega_{\bullet e} - \omega) \right] \tag{3.19}$$

By identifying δ in (3.14) with the corresponding expression in (3.19) we find from (3.14) that we have an instability if $\omega < \omega_{\bullet e}$. We find:

$$\mathrm{Im}\omega = \frac{m_e \nu_{ei}}{k_{\|}{}^2 T_e} \omega_{\bullet e}(\omega_{\bullet e} - \omega) \tag{3.20}$$

Since, however, we have Re $\omega < \omega_{\bullet e}$, from (3.6b) we realize that we need some additional effect in order to have an instability. Since the ions are not so strongly influenced by collisions with electrons we use (3.10) with $k_{\|} = 0$ for the ion density perturbation. Combining this equation with (3.19) we find the dispersion relation:

$$\omega\left(1 + k_y{}^2 \rho_s{}^2\right) = \omega_{\bullet e} + \nu_{ei} \frac{m_e}{k_{\|}{}^2 T_e} \omega_{\bullet e}(\omega_{\bullet e} - \omega) = 0 \tag{3.21}$$

Writing the solution as $\omega = \omega_r + i\gamma$ where $\gamma << \omega_r$ we find

$$\omega_r \approx \omega_{\bullet e}\left(1 - k_y{}^2 \rho_s{}^2\right) \tag{3.22a}$$

$$\gamma = \nu_{ei} \frac{m_e}{k_{\|}{}^2 T_e} \omega_{\bullet e}{}^2 k_y{}^2 \rho_s{}^2 \tag{3.22b}$$

We see from (3.22b) that the ion inertia, $k_y \rho_s$, is essential for an instability to develop. We may explain the instability in the way that the ion inertia causes the particle drifts of electrons and ions in the perpendicular plane to become different. This leads to charge separation effects if we have a density perturbation and due to the electron-ion collision the electrons are not able to instantly neutralize the charge separation by moving along the magnetic field.

3.3 MHD Type Modes

As mentioned in the beginning of this chapter there are two classes of low frequency modes, drift modes and MHD type modes. While the drift modes are characterized by essentially free electron motion along the field lines leading to the Boltzmann distribution (3.3b) or minor modifications thereof, the MHD type modes are modes where the parallel electric field to lowest order vanishes. This can in the electrostatic case be accomplished by a very small $k_{\|}$ ($\omega >> k_{\|} v_{the}$) and in the electromagnetic case by a cancellation between electrostatic and induction parts of E. In both cases the parallel electron motion is strongly impeded and as a consequence new types of instabilities may arise. These unstable modes may be divided into two classes: *pressure driven modes*, here represented by interchange and ballooning modes, and *current driven modes*, here represented by kink modes. The transition between MHD and drift type modes in a simple case is shown by Exercise 3.8.

3.3.1 Alfve´n Waves

As an example of a simple two fluid derivation of an MHD mode we will now show the derivation of Alfve'n waves. This is the most fundamental MHD type eigenmode and Alfve'n waves were originally called Electromagnetic hydrodynamic waves. We start from the quasineutrality condition

$$\nabla \cdot \mathbf{j} = 0 \tag{3.23a}$$

Since $\nabla \cdot (\mathrm{n}\mathbf{v}_*) = 0$ and the E × B drifts are equal, only the polarisation drift contributes to the perpendicular current. Thus (3.23a) becomes:

$$\nabla \cdot (n\mathbf{v_p}) = -\nabla \cdot \mathrm{j}_{||} = \frac{1}{\mu_0}\Delta \mathrm{A}_{||} \tag{3.23b}$$

Where we used

$$\delta B = \nabla x(\mathrm{A}_{||}\hat{\mathbf{z}})$$

Now the MHD constraint

$$E_{||} = 0 \tag{3.24a}$$

or

$$\mathrm{A}_{||} = \frac{1}{\omega}\phi \tag{3.24b}$$

can be used to express (3.23b) in only the potential. We then get the dispersion relation for Alfve'n waves:

$$\omega^2 = \mathrm{k}_{||}{}^2 \mathrm{v_A}{}^2 \tag{3.25}$$

where $\mathrm{v_A} = \frac{\mathrm{B_0}}{\sqrt{\mu_0 \mathrm{n m_i}}}$ is the Alfve'n velocity.

This derivation is probably simpler than with the one fluid equations.

3.3.2 Interchange Modes

One of the most dangerous modes in fusion machines is the Interchange mode, sometimes also called flute mode, which tends to interchange "flux tubes" of different pressure, thus causing a convective transport. (Compare also the section "Interchange modes analyzed by the energy principle method"). Interchange modes are unstable when the magnetic curvature generates a centrifugal force, due to thermal motion along the field lines, which is directed in the opposite direction of

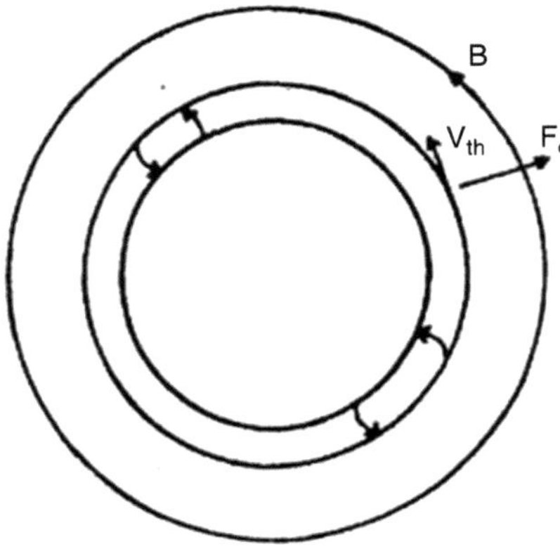

Fig. 3.8 Interchange of fluid elements in a cylinder

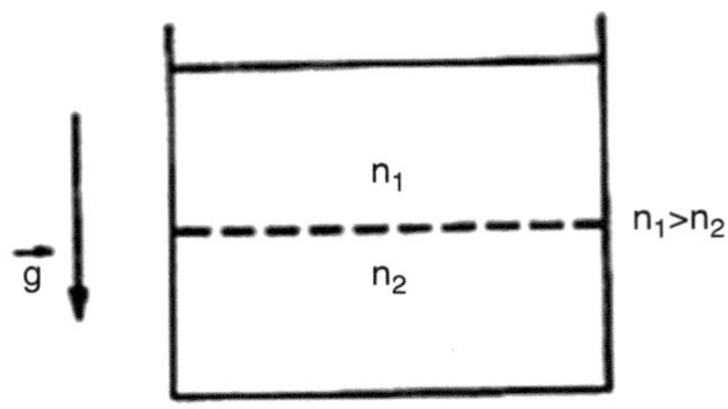

Fig. 3.9 Rayleigh-Taylor instability

the pressure gradient. As a simple example we consider a z-pinch with only poloidal magnetic field. The figure shows fluid elements that would tend to change places. A simple fluid analogue of this instability is the Rayleigh-Taylor instability when a heavy fluid is resting on a light fluid. The density gradient here corresponds to the pressure gradient for the interchange mode while the gravity represents the centrifugal force (Figs. 3.8, 3.9).

The gravity may thus be used to simulate a curvature and this is the main reason why we included it in Eq. (2.11a). We will here neglect finite Larmor radius effects that would correspond to diamagnetic drift contributions to the polarization drift and stress tensor drifts. We will also make the approximation $k_\parallel = 0$ (flute mode). This is the most unstable mode since a mode with $k_\parallel \neq 0$ would tend to bend the frozen in magnetic field lines, thus increasing the magnetic energy. We may obtain a dispersion relation by substituting the drifts into the low frequency condition:

$$\nabla \cdot \mathbf{j} = 0$$

In the present case this gives

$$\nabla \cdot \left[en(\mathbf{v_{pi}} + \mathbf{v_{gi}} - \mathbf{v_{ge}})\right] = 0$$

A linearization, using again $k_\perp << \kappa$, leads to:

$$n_0 \nabla \cdot \left[-\frac{1}{B_0 \Omega_c} \left(\frac{\partial}{\partial t} + \mathbf{v_{gi}} \cdot \nabla \right) \nabla_\perp \phi \right] + (\mathbf{v_{gi}} - \mathbf{v_{ge}}) \cdot \nabla \delta n = 0 \tag{3.26}$$

The density perturbation can here be obtained from the electron continuity equation. Note that this is a special case! We consider v_{ge} to be small. We may then ignore it in the continuity equation:

$$\frac{\partial \delta n_e}{\partial t} + \mathbf{v}_E \cdot \nabla n_0 = 0$$

Or

$$\frac{\delta n_e}{n_0} = \frac{\omega_{\bullet e}}{\omega} \frac{e\phi}{T_e} \tag{3.27a}$$

Substituting (3.25) into (3.24) we obtain the dispersion relation:

$$\omega(\omega - k_y v_{gi}) = -\kappa \left(g_i + \frac{m_e}{m_i} g_e \right) \frac{k_y^2}{k_\perp^2} \tag{3.27b}$$

where $\kappa = -d\ln(n)/dx$. We note here that if the gravity is replaced by a real curvature $\vec{\kappa} = -d\ln(P)/dx$. We here easily recognize the part $\omega^2 = -\kappa\ g$ corresponding to the Rayleigh-Taylor instability. When g_j is due to curvature we have $g_i = 2\ T_i/m_i\ R_c$ where R_c is the radius of curvature. Then the dispersion relation may be rewritten as

$$\omega(\omega - k_y v_{gi}) = -\frac{2\kappa(T_e + T_i)}{m_i R_c} \frac{k_y^2}{k_\perp^2} \tag{3.28}$$

The drift $k_y v_{gi}$ here is stabilizing. This means that modes with small k_y are the most unstable modes. As it turns out the lowest order FLR correction has the same influence but is typically larger than the drift term kept here. In the fluid description this instability is due to the density gradient in a very simple way, i.e. the fluid motion happens in the direction of the gradient. In a plasma the convection is also in the direction of the pressure gradient but the actual physical process is more complicated since in a plasma forces in the perpendicular plane primarily give rise to motion perpendicular to the force. The source of the instability is the difference in gravity (curvature) drifts of electrons and ions which in combination with a density perturbation leads to a charge separation. When $k_\parallel = 0$ the electrons cannot short circuit this charge separation which leads to an electric field that is perpendicular to the pressure gradient and the magnetic field. When the pressure gradient and the gravity have opposite directions this electric field causes an $E \times B$ drift which enhances the original perturbation (Fig. 3.10).

The real frequency caused by $k_y v_{gi}$ is stabilizing since it changes the polarity of the field. A gravity due to field curvature is shown in Fig. 3.11.

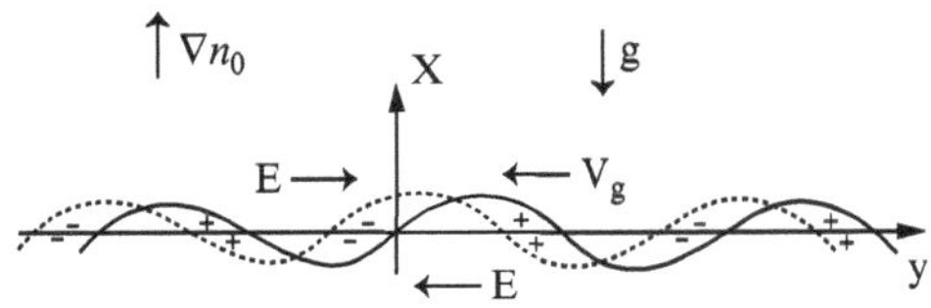

Fig. 3.10 The mechanism of interchange instability

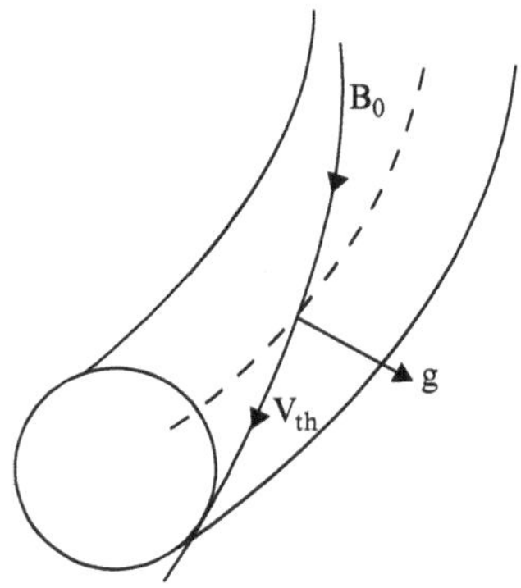

Fig. 3.11 Gravity representing a centrifugal force

3.3.3 The Convective Cell Mode

If we let $\kappa \to 0$ in (3.27) the two branches become uncoupled and we have one mode with $\omega = 0$ and one with $\omega = k_y v_{gi}$ {The mode $\omega = 0$ corresponds to a stationary convection (compare Fig. 3.3) and is called the convective cell mode}. When finite ion Larmor radius effects are included, the ion diamagnetic drift frequency is added to v_{gi}. This means that we have two different modes also if we let $R_c \to \infty$. In a real system with curvature and density gradient the convective cell thus turns into the interchange mode. Since it has no variation along the field lines it will experience the average curvature along the field line. This curvature is usually favourable in a tokamak leading to a real eigenfrequency. This will tend to reduce the transport (compare Chap. 9).

3.3.4 Electromagnetic Interchange Modes

In a physical system with magnetic shear (see Sect. 6.2) the approximation $k_\parallel = 0$ cannot be exactly fulfilled since the mode has a finite extension in space and since the magnetic field direction is space dependent. Another situation may be when the average curvature is stabilizing but there are local regions along a field line where the curvature is destabilizing. In order to see qualitatively what the consequences of a finite $k_\parallel$ will be, we shall here simply include a finite $k_\parallel$ into our simple slab geometry. In this case our previous description for the perpendicular motion

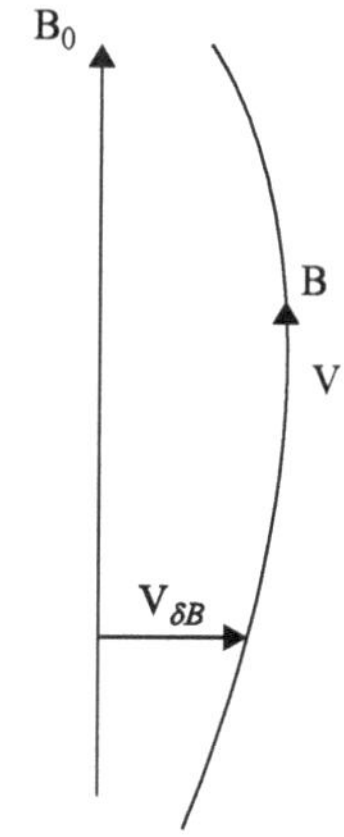

Fig. 3.12 The magnetic energy increases when a field line bends. This effect is stabilizing

continues to hold. For the parallel direction we may neglect ion motion assuming $\omega >> k_{\|}\, c_s$ compare (3.11). The parallel electron motion can in the simplest case be described in the same way as for shear Alfve'n waves, i.e. we combine the Ampe're law along the field lines (2.18) with the perfect conductivity condition (2.19). Then using (2.17) in the form

$$\nabla \cdot \mathbf{j}_{\perp} = -\nabla \cdot \mathbf{j}_{\|}$$

we arrive at the dispersion relation

$$\omega(\omega - k_y \mathrm{v}_{\mathrm{gi}}) - k_{\|}{}^2 \mathrm{v}_{\mathrm{A}}{}^2 = -\frac{2\kappa(T_e + T_i)}{m_i R_c} \frac{k_y{}^2}{k_{\perp}{}^2} \tag{3.29}$$

The new effect here is the bending of the field lines, represented by the Alfv'en frequency. This effect is stabilizing since the line bending increases the magnetic energy as (Fig 3.12). For small k_y the dispersion relation (3.28) leads to a pressure balance condition for stability. In a torus with periodic curvature and unfavourable curvature regions of length $L_c \approx 2\pi q R_c$. where q is the safety factor (compare the chapter on toroidal mode structure) we may to order of magnitude take $k_{\|} \approx 1/qR$ where R is the large radius and $\kappa \approx 1/a$ where a is the small radius. The pressure balance condition for stability then takes the simple form

$$\beta \leq \beta_c = \frac{a}{q^2 R} \tag{3.30}$$

where $\beta = 2\mu_0 n\ (Ti + Te)/B^2$ is the ratio of plasma and magnetic field pressure (or energy density). This β limit is typical of ballooning modes in toroidal machines. These modes are interchange modes localized in regions of unfavourable curvature and are one of the main limiting instabilities for the achievable β in tokamaks (Fig. 3.13).

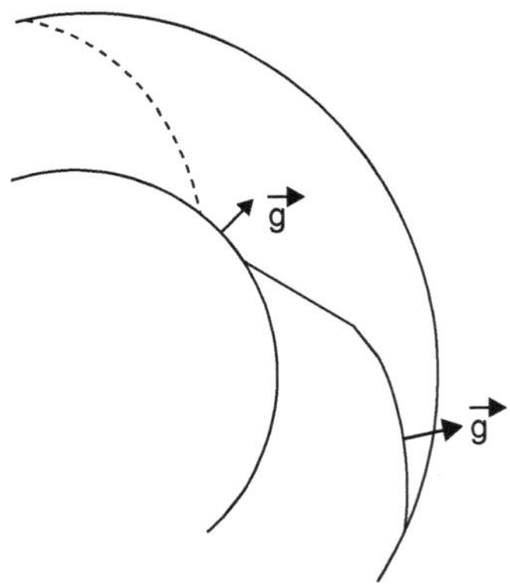

Fig. 3.13 Field line curvature

Another source of finite $k_{||}$ is the radial extent of a mode in a system with shear. We may here think of the previous slab quantities as averaged over the mode profile. The averaged curvature may be written:

$$\left\langle \frac{\kappa}{R_c} \right\rangle \approx_c - \frac{\delta}{R} \frac{1}{\mathrm{P}} \frac{dP}{dr}$$

where δ is a factor due to averaging and we introduced the radial coordinate r instead of x.

The averaging of $k_{||}$ leads to (compare Chap. 6).

$$<k_{||}> \approx \Delta' s / qR$$

where s = dlnq/dlnr and r is the radial coordinate. We then obtain the stability condition:

$$(\Delta' s)^2 > - \delta \beta q^2 R \frac{d \ln p}{dr} \tag{3.31a}$$

As it turns out (a discussion of this result will be given later) (Fig. 3.13).

$$\Delta' \approx \frac{1}{2}$$

For a z-pinch where the magnetic field is purely poloidal the problem becomes singular since q = 0. If we, however, treat q as small but finite we obtain $\delta = -r/Rq^2$ and the q dependence disappears. We then obtain the *Suydam criterion* [2]

$$\frac{1}{4} s^2 + r \frac{d\beta}{dr} > 0 \tag{3.31b}$$

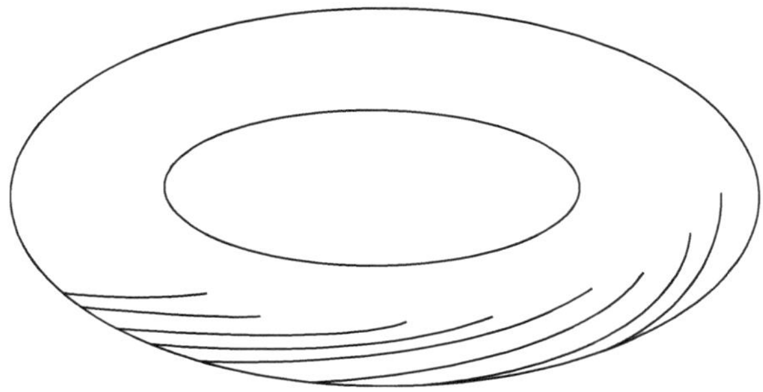

Fig. 3.14 Ballooning mode perturbations on a torus

For a torus with both poloidal and toroidal magnetic field it can be shown that $\delta = (r/R)(1 - 1/q^2)$ and the stability condition turns into the *Mercier criterion* [2] (Fig. 3.14).

$$\frac{1}{4}s^2 + r\frac{d\beta}{dr}(1 - q^2) > 0 \tag{3.32}$$

This means that the average curvature is stabilizing for $q > 1$ and instability is only possible for $q < 1$. The Mercier criterion holds only for modes that are highly elongated along the field lines and experience only the average curvature. When localized ballooning modes are taken into account the possibility for instability increases strongly and a rather typical β limit is $\beta_c/2$.

3.3.5 *Kink Modes*

One of the most dangerous instabilities in current carrying cylindrical and toroidal plasmas is the kink instability. It corresponds to a bending of the whole system (global mode) so that the change in magnetic pressure tends to increase the perturbation (see Fig. 3.15).

Although this mode is most easily visualized for global perturbations in combination with sharp current boundaries, the only necessary ingredient is a background current gradient perpendicular to the magnetic field.

We will here include the kink mode in our previous analysis by including a background current with a gradient in the x (radial direction). If we neglect the associated frequency shift ($\omega >> k_{||}v_{||}$) the only new terms we need to include are the $\mathbf{v}_{\delta\mathbf{B}}$ drifts in (2.11h) for electrons and ions. This leads to a new contribution to div $\mathbf{J}_\perp$ as

$$n_0\nabla \cdot \left[e(n_i v_{||0i} - n_e v_{||0e})\frac{\delta\mathbf{B}_\perp}{B_0}\right] = \frac{\delta\mathbf{B}_\perp}{B_0} \cdot \nabla\mathbf{J}_{||0} \tag{3.33}$$

where $J_{||0}$ is the background current. Keeping also our previous driving pressure term we may write the equation div $\mathbf{j} = 0$ as

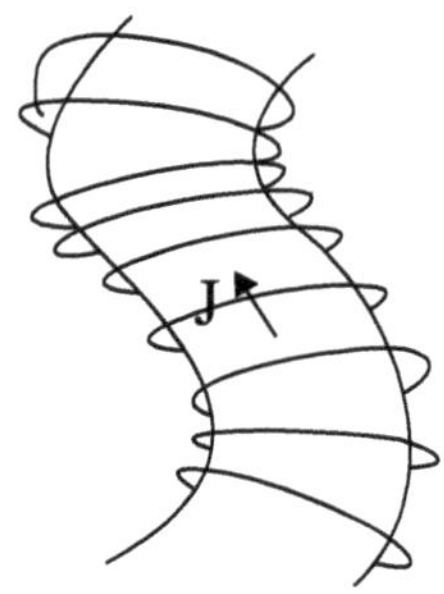

Fig. 3.15 Kink perturbation

$$-\frac{1}{\mu_0}\hat{\mathbf{z}}\cdot\nabla\Delta_\perp A_{||} = -\nabla_\perp\cdot\left[en\mathbf{v}_{\rm pi} + en(\mathbf{v_{gi}} - \mathbf{v_{ge}}) + J_{||0}\frac{\delta\mathbf{B}_\perp}{\mathrm{B}_0}\right] \tag{3.34}$$

where $A_\| = -(i/\omega)\acute{\mathbf{z}}\bullet\mathrm{grad}\varphi$ according to the condition that $E_{||} = 0$. Proceeding as before but now neglecting for simplicity also the frequency shift due to the ion gravity drift we obtain the dispersion relation for $k_\perp >> |\,\mathrm{grad}\ \ln n\,|$:

$$\omega^2 = k_{||}{}^2 \mathrm{v_A}{}^2 - \frac{2\kappa(T_e + T_i)}{m_i R_c}\frac{k_y{}^2}{k_\perp{}^2} + \frac{\mathrm{B}_0}{\mathrm{n}_0 m_i}\frac{k_{||}k_y}{k_\perp}\frac{dJ_{||0}}{dx} \tag{3.35}$$

Since div $\delta\mathbf{B}_\perp = 0$ the div• has to operate on the background quantity $J_{\|0}$ in the last term of (3.35).

This has the consequence that the kink term usually is small for local modes since also $k_\|$ is small. As it turns out, however, the only new term that arises if we relax the condition $k_\perp >> |\,\mathrm{grad}\ \ln n\,|, |\,\mathrm{grad}\ \ln J_{||0}|$ is the density gradient contribution from the polarization drift. This effect is usually neglected for global modes so that (3.35) can in fact be written as an eigenvalue equation for such modes. The kink term may also become important locally if the background current gradient is locally large. With $\mathrm{grad}(\ln n_0) \approx -1/a$ where a is the small radius, the β limit (3.29) is modified to:

$$\beta \le \frac{a}{q^2 R} + \frac{\mu_0 a}{B_0 q k_\perp}\frac{dJ_{||0}}{dx} \tag{3.36}$$

where we used $k_{||} \approx 1/qR$. The condition (3.36) shows that the kink term decreases the β limit (destabilizing) if $d\,J_{||0}/dx < 0$ which is the typical case. If we write $d\,J_{||0}/dx = -\kappa_b en\,\mathrm{v}_{||0}$ we obtain the β limit in the form:

$$\beta \le \frac{a}{q^2 R} - \frac{\Omega_{ci}\mathrm{v}_{||0}}{k_\perp \mathrm{v_A}{}^2}\kappa_b a \tag{3.37}$$

For the kink term to change the β limit appreciably we need $\kappa_b a = 5$ for typical tokamak parameters if $k_\perp \mathrm{v_A} \approx \Omega_{ci}$.

3.3.6 *Stabilization of Electrostatic Interchange Modes by Parallel Electron Motion*

In the preceding section we have found that interchange modes may be unstable for zero $k_{||}$ or when $\kappa g > k_{||}^2 \, v_A^2$. In the opposite case when $k_{||}v_A$ becomes large the mode becomes electrostatic. (This will come out of the kinetic treatment in Chap. 4 but can also easily be seen from the fluid equations as shown by exercise 8).

In the electrostatic limit when $\omega << k_{||}v_{the}$ the electrons are Boltzmann distributed, i.e.

$$\frac{\delta n_e}{n_0} = \frac{e\phi}{T_e} \tag{3.38}$$

This relation comes out of the parallel equation of motion and is not influenced by gravitation. For the ions we may use (3.29) where the gravitation introduces a doppler shift, i.e. $\omega \to \omega - k_y v_{gi}$ and where we neglect $k_{||}^2 c_s^2$.

$$\frac{\delta n_i}{n_0} = \left(\frac{\omega_{\bullet e}}{\omega - k_y v_{gi}} - k_y^2 \rho_s^2 \right) \frac{e\phi}{T_e} \tag{3.39}$$

Then using quasineutrality we obtain the dispersion relation

$$\omega = \frac{\omega_{\bullet e}}{\omega + k_y^2 \rho_s^2} - k_y v_{gi} \tag{3.40}$$

This is just the dispersion relation for an ordinary drift wave where a frequency shift $k_y v_{gi}$ has been added. Thus there is no instability. The reason for this is that the electrons are free to move along the field lines to cancel space charge. In the electromagnetic case the electron motion along the field lines is impeded by magnetic induction, thus providing the necessary conditions for instability.

3.3.7 *FLR Stabilization of Interchange Modes*

As it turns out the lowest order FLR effect is often significant and introduces qualitatively new effects while the higher order effects usually only modify previously known results. We shall here demonstrate the stabilizing influence of FLR effects on interchange modes. This is now very easily done by just replacing div $(n v_{pi})$ with (see Chap. 2)

$$\nabla \cdot \left[n(\mathbf{v_p} + \mathbf{v}_\pi) \right] = -ink^2 \rho^2 (\omega - \omega_{*iT}) \frac{e\phi}{T_e} \tag{2.37}$$

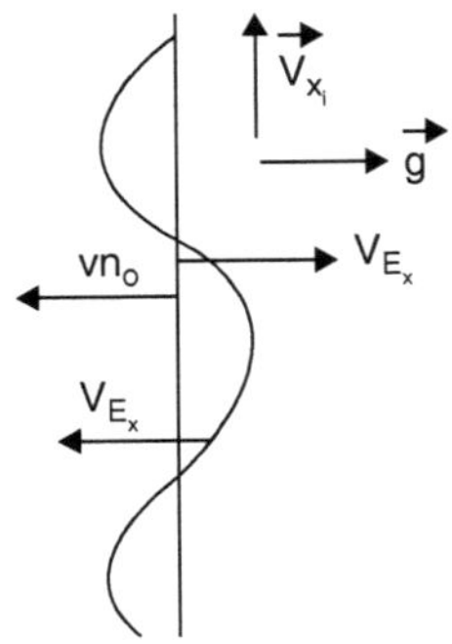

Fig. 3.16 The ion diamagnetic drift causes a propagation which changes the polarity of the charge separation

This result can be verified by using the stress tensor $v_{\pi i}$. This means that the lowest order FLR term can be obtained by just shifting ω by $\omega_{\bullet i}$ in the ion polarization drift. This effect is, according to conventional estimates, larger than the shift due to the gravity drift in (3.26) due to the following estimate. When v_{gi} is due to curvature and gradB drifts it may be written:

$$\left|v_{gi}\right| = \frac{2T_i}{mR_c\Omega_{ci}} = \frac{\rho_i}{R_c} v_{thi} \tag{3.41}$$

The diamagnetic drift may be written:

$$\left|v_{\bullet i}\right| = \frac{T_i\kappa}{mR_c\Omega_{ci}} \approx \frac{1}{2}\kappa\rho_i v_{thi} \tag{3.42}$$

Now R_c is typically approximated by the large radius R in a torus and κ is on the average over the profile 1/a where a is the small radius. We thus arrive at the estimate

$$\left|\frac{v_{gi}}{v_{\bullet i}}\right| \approx 2\frac{a}{R} \tag{3.43}$$

When a << R we may thus consider the gravity drift to be small (this is no longer fulfilled for the newest generation of large tokamaks). As will be pointed out later, κ is typically much smaller than 1/a in the interior of tokamaks so the magnetic drift should usually not be considered to be small. However, this leads to major complications for using fluid models and particular advanced closures have to be used. Thus for the time being we will consider the magnetic drift to be small.

The dispersion relation then takes the form (Fig. 3.16):

$$\omega(\omega - \omega_{\bullet iT}) - k_{\|}^2 v_A^2 = -\frac{2\kappa(T_e + T_i)}{m_i R_c}\frac{k_y^2}{k_\perp^2} \tag{3.44}$$

With the solution

$$\omega = \frac{1}{2}\omega_{\bullet iT} \pm \sqrt{\frac{1}{4}\omega_{\bullet iT} + k_{\|}{}^2 \mathrm{v_A}^2 - \frac{2\kappa(T_e + T_i)}{m_i R}\frac{k_y{}^2}{k_\perp{}^2}} \tag{3.45}$$

The stabilizing FLR term changes the β limit as given by (3.29) to

$$\beta \leq \frac{a}{q^2 R} + \frac{1}{4}\omega_{\bullet iT}{}^2 aR/\mathrm{v_A}^2 \tag{3.46}$$

This is an important effect for modes with large k_y. In a tokamak there are, however, unstable modes with so small k_y that FLR effects can be neglected within the present description. When curvature effects that are higher order in a/R and geometrical effects in D-shaped tokamaks are included, however, it appears that finite Larmor radius effects have to be included as well. In addition to the stabilizing effect we also notice from (3.45) that the mode now has a finite real part of the eigenfrequency also at marginal stability and in the unstable case. This real part of ω corresponds to propagation in the y direction and is in fact the reason also for the stabilizing effect.

The interchange of fluid elements by convection, which is the fundamental instability we are considering, is a fluid motion and the fluid only moves in the x direction (if we neglect k_x). The perturbation, however, propagates in the y direction, causing a change in the direction of convection after a time $\tau = 2\,\pi/\omega_{\bullet iT}$. We then realize that if this time is short as compared to the time needed for the instability to develop it will become stabilized. The physical interpretation of the FLR effect is a modification of the E × B drift due to the inhomogeneity of **E** along the gyro orbit. This effect is usually only important for ions and leads to a charge separation in the presence of a density gradient. This charge separation will have different sign in region where $\mathbf{v}_E$ is directed in the positive or negative x direction and leads to a propagation of the perturbation in the y direction.

3.3.8 Kinetic Alfve'n Waves

In order to exemplify the modification of MHD type modes in the presence of a finite $E_\|$ we will now study a mode which is of a particular interest for the heating of fusion plasmas, the kinetic Alfve'n wave. For simplicity we will here take the limit of a homogeneous plasma and assume that $T_i << T_e$. We will start by generalizing the Boltzmann distribution (3.3) to the electromagnetic case. This is straightforward by just adding $(e/m)dA_\|/dt$ to the right-hand side of (3.2). This leads to

$$\frac{\delta \mathrm{n_e}}{\mathrm{n_0}} = \frac{e\phi}{\mathrm{T_e}} - \frac{\omega}{k_\|}\frac{eA_\|}{T_e} \tag{3.47}$$

In the case of a homogeneous plasma we may completely neglect perpendicular electron motion. The electron continuity equation may then be written

$$\frac{\partial n_e}{\partial t} - \frac{1}{e}\nabla \cdot \mathbf{j}_{||\mathbf{e}} = 0 \tag{3.48a}$$

Now, neglecting parallel ion motion ($\omega >> k_{||}c_s$) we may write $j_{||e} \approx j_{||}$ and make use of (3.23b). This leads to

$$\frac{\delta n_e}{n} = -\frac{k_{||}k_\perp^2}{\omega}\frac{1}{\mu_0 en}A_{||}$$

Since

$$\frac{1}{\mu_0 en} = \frac{B^2}{\mu_0 nm_i}\frac{m_i}{eB^2} = \frac{v_A^2}{\Omega_{ci}B} = \rho_s^2 v_A^2 \frac{e}{T_e}$$

We then obtain

$$\frac{\delta n_e}{n} = -k_\perp^2 \rho_s^2 \frac{k_{||}}{\omega} v_A^2 \frac{eA_{||}}{T_e} \tag{3.48b}$$

Combining (3.47) and (3.48b) we obtain:

$$A_{||} = \frac{k_{||}/\omega}{1 - k_\perp^2 \rho_s^2 k_{||}^2 v_A^2/\omega^2}\phi$$

Using this result instead of (3.24) in the derivation of (3.25) we obtain the dispersion relation of the kinetic Alfve'n wave

$$\omega^2 = k_{||}^2 v_A^2 (1 + k_\perp^2 \rho_s^2) \tag{3.49}$$

This is the most simple form of the kinetic Alfve'n wave. The heating properties of this mode are related to singularities in the mode structure in plasmas with magnetic shear. The origin of such singularities is discussed in the section of kink modes in Chap. 6. As is easily verified the $k^2\rho^2$ part of (3.49) is due to a parallel electric field which may be expressed as

$$E_{||} = i\frac{k^2\rho^2 k_{||}^2 v_A^2}{\omega^2 - k_\perp^2 \rho_s^2 k_{||}^2 v_A^2} k_{||}\phi$$

These results can easily be generalized to inhomogeneous plasmas as we will soon see (Exercise 8).

3.4 Quasilinear Diffusion

We shall now consider the particle transport due to low frequency modes in magnetized plasmas. We start by observing the correspondence between the continuity equation and the diffusion equation, i.e.

$$\frac{\partial n}{\partial t} + \nabla \cdot (n\mathbf{v}) = 0$$

may be written

$$\frac{\partial n}{\partial t} = -\nabla \cdot (\Gamma) \tag{3.50}$$

where Γ is the flux which, according to Fick's law fulfills

$$\Gamma = -\ D\nabla n \tag{3.51}$$

where D is the diffusion coefficient and (3.50) reduces to the diffusion equation

$$\frac{\partial n}{\partial t} = \nabla \cdot (D\nabla n) \tag{3.52}$$

Equation 3.52 is only of interest in so far as it describes a secular steady state diffusion. We thus want to average (3.52) over the harmonic time and space variation of the fluctuations. In an inhomogeneous plasma a harmonic wave will always obtain a superimposed slow space variation of the amplitude due to the inhomogeneity, i.e.

$$\phi = \widehat{\phi}(x)e^{-i(\omega - \mathbf{k}\cdot\mathbf{r})} + c.c \tag{3.53}$$

Where the inhomogeneity is in the x direction. The flux in the x direction averaged over the harmonic variation is now:

$$< \Gamma_x > = \sum \delta n_k v_{kx}{}^* + c.c. \tag{3.54}$$

Where

$$v_{kx} = v_{Ex} = -i\frac{k_y}{B}\phi_k$$

The electrons are assumed to be close to Boltzmann distributed but with a small imaginary correction due to dissipative effects, i.e.

$$\frac{\delta n_e}{n_0} = (1 - i\delta)\frac{e\phi}{T_e} \tag{3.55}$$

We may then rewrite (3.54) as:

$$< \Gamma_x > = 2n_0 \frac{T_e}{eB} \sum \left| \frac{e\phi_k}{T_e} \right|^2 k_y \delta_k \tag{3.56}$$

or, as a result of (3.51)

$$D_e = \frac{2T_e}{\kappa eB} \sum \left| \frac{e\phi_k}{T_e} \right|^2 k_y \delta_k \tag{3.57}$$

$$\kappa = -\frac{1}{n_0} \frac{dn_0}{dx}$$

We notice that the diffusion is due to the imaginary part of the deviation from the Boltzmann distribution. This dependence is such that unstable waves cause diffusion in positive x, i.e. towards the plasma boundary. As it turns out δ_k will be proportional to κ in most cases of practical interest so that D remains finite when $\kappa \to 0$. It is also interesting to consider the ion diffusion. In the simpliest case with only $E \times B$ convection the ion density perturbation is:

$$\frac{\delta n_{ik}}{n_0} = \frac{\omega_{\bullet e}}{\omega_k} \frac{e\phi_k}{T_e} \tag{3.58}$$

in the region $\omega_k >> k_{||} c_s, k_\perp^2 \rho_s^2 << 1$. Equation 3.54 then becomes

$$< \Gamma_x > = n_0 \frac{T_e}{eB} \sum \left| \frac{e\phi_k}{T_e} \right|^2 ik_y \frac{\omega_{\bullet e}}{\omega_k} + c.c.$$

Now writing $\omega_k = \omega_r + i\gamma_k$ we obtain

$$< \Gamma_x > = 2n_0 \frac{T_e}{eB} \sum \left| \frac{e\phi_k}{T_e} \right|^2 \frac{k_y \gamma_k}{\omega_r^2 + \gamma_k^2} \omega_{\bullet e}$$

and

$$D_i = 2\frac{T_e}{eB} \sum k_y \rho_s \frac{\gamma_k k_y c_s}{\omega_r^2 + \gamma_k^2} \left| \frac{e\phi_k}{T_e} \right|^2 \tag{3.59}$$

The instability saturates when κ has decreased to zero due to the diffusion. Another possible mechanism for saturation is when the nonlinear contribution to the radial derivative of the convective density perturbation becomes comparable to that of the linear i.e. when $k_x \delta n = \kappa n_0$ or

$$\frac{\delta n_{ik}}{n_0} = \frac{1}{k_x L_n} \tag{3.60}$$

where $L_n = 1/\kappa$ is the density inhomogeneity scale length. This is reasonable since the radial (x) inhomogeneity is driving the instability. This estimate is called the mixing length estimate. It typically leads to a turbulence level of a few percent. Since the linear perturbation is

$$\delta n = -\xi \cdot \nabla n_0 = -\xi_x \frac{dn_0}{dx} = \xi_x \frac{n_0}{L_n} \tag{3.61}$$

And

$$v_{kx} = -i\omega\xi_x$$

We obtain from (3.54)

$$< \Gamma_x > = \frac{n_0}{L_n} \sum -i\omega_k |\xi_x|^2 + c.c. \tag{3.62}$$

Now combining (3.60) and (3.61) we obtain

$$k_x |\xi_x| \leq 1 \tag{3.63}$$

This means that the displacement is a sizeable fraction of a wavelength. Now, using (3.63) in (3.62) we obtain the estimate

$$D_i = \Gamma \frac{L_n}{n_0} \leq 2 \sum \gamma_k / {k_x}^2$$

If we interpret k_x as a correlation length of the full space variation we omit the summation. The estimate is then actually written in the form

$$D_i \approx \gamma_k / {k_x}^2 \tag{3.64}$$

This result can also be obtained by renormalization (Dupree 1987), When the dominant nonlinearity is of the E × B convective type as in the continuity or energy equation

$$\frac{\partial n}{\partial t} = v_{Ex} \frac{\partial}{\partial x} n$$

we can estimate the saturation level by balancing the linear growth, i.e. $\partial/\partial t \rightarrow \gamma$ with the nonlinearity. Then, representing "grad" by the inverse space scale of the full perturbation, the density (or temperature) perturbation cancels and we obtain the saturation level [44, 49].

$$\frac{e\phi}{T_e} \approx \frac{1}{k_x \rho_s} \frac{\gamma}{k_y c_s} \tag{3.65}$$

Equation 3.63 is then replaced by

$$|\xi_x| \leq \frac{\gamma}{|\omega|} \frac{1}{\mathrm{k_x}} \tag{3.66}$$

Equation 3.66 shows that a real eigenfrequency reduces the step length because the convection oscillates in time. Using (3.66) in (3.62) we now obtain [44]:

$$\mathrm{D} \approx \frac{\gamma^3/{k_x}^2}{{\omega_r}^2 + \gamma^2} \tag{3.67}$$

which turns into the mixing length estimate [3, 10] when $\gamma \gg \omega_r$ While (3.64) has the character of upper limit, (3.67) is a more direct estimate of the transport level. Recently a more general derivation has been made of (3.67) from a non-Markovian Fokker Planck equation [3.48, 3.49]. It shows that (3.67) is a quite general expression, which only lacks off diagonal elements. In the general expression, ω_r also contains the non-linear frequency shift.

3.5 Confinement Time

It is important to relate the diffusion coefficient to the confinement time, τ. The confinement time is defined as the characteristic time for the decrease in total number of particles N or total energy due to diffusion. For particle diffusion from a cylinder of radius r and length L we have

$$\tau = \frac{N}{dN/dt} = \frac{n\pi r^2 L}{2\pi r L \Gamma} = \frac{nr}{2\Gamma} \tag{3.68}$$

Where n is the particle density and Γ is the particle flux given by (3.51). We then obtain

$$\tau = \frac{rL_n}{D} \tag{3.69}$$

where $L_n = 1/\kappa$ is the characteristic length scale of density variation. If we take $L_n = r$ we find that for classical diffusion $\tau \sim r^2B^2$ which would mean that by increasing the magnetic field we can build a smaller machine obtaining the same confinement time. For quasilinear diffusion two cases are usually discussed. These are Gyro Bohm diffusion where D will scale as B^{-2} and Bohm diffusion where $D \sim B^{-1}$. For Bohm diffusion a stronger increase in the magnetic field is necessary for a reduction in size. Fortunately Gyro Bohm diffusion has been found to dominate in the core. An increase in B also allows higher confined pressure and density.

Clearly we can make the same derivation of the energy confinement time τ_E in terms of the thermal conductivity χ. As is evident from (1.8) the dependencies of τ_E on a and R are much more complicated in a real toroidal system. The scaling with R seems to be, at least partly, due to the curvature radius of curvature driven modes. It is, in fact, possible to obtain the scaling $R^{1.5}$ from (3.67) when the real eigenfrequency dominates and the growthrate is given by the root of κg .

3.6 Discussion

We have in the present chapter studied new eigenmodes associated with the inhomogeneity of a plasma. These modes are very fundamental since they are the plasmas response to an inhomogeneity. They will, accordingly, have the effect to cause anomalous transport that tends to reduce the driving inhomogeneity as also demonstrated. The modes studied were either of an MHD type with no parallel electric field or of a drift type with electrostatic, Boltzmann electrons. As seen from Exercise 8, where a transition between these two types is made, the MHD type modes are more global. They also generally have larger growth-rates. In this chapter we have used a simple slab geometry to show the most fundamental properties of the modes. In Chap. 6 more realistic geometries will be introduced. We will first in Chap. 4 use a kinetic theory to rederive dispersion relations for modes studied here.

Exercises

1. Explain why $\mathbf{v}_*$ does not contribute to (3.9).
2. (a) Generalize the derivation of (3.6) to the case of finite k_x.
   (b) Do the same with (3.10).
3. Discuss which of the effects included in (3.10) that corresponds to compressibility.
4. An effect of finite Larmor radius (FLR) is that the ion particle E × B drift is reduced to

$$v_{Ei} = \frac{1}{B}(\hat{\mathbf{z}} \times \nabla\phi_k)\left(1 - \frac{1}{2}k^2\rho_i\right)$$

   As explained in the section "Interpretation of drifts" in Chap. 2 we are allowed to replace fluid drifts by particle drifts throughout in the equation div $\mathbf{j} = 0$. Use the FLR corrected ion E × B drift to derive modified versions of the dispersion relations (3.28) and (3.29).
5. Show that we by adding electron-ion collisions in the same way as in (3.17) can obtain an instability driven by gravity in (3.40), i.e. for $\omega << k_\perp v_{the}$. This is a resistive interchange mode. The derivation may be simplified by assuming that $|k_y v_g| < \omega_{*e}$ holds for both electrons and ions.

6. Assume a simple cylindrical geometry where the magnetic field is in the θ direction and where the pressure decreases in the radial direction. Derive the β limit as a function of mode number m, $\varphi \sim e^{im\theta}$ when we make the replacement $\kappa \rightarrow (1/P)dP/dr$.
7. In a torus with both toroidal and poloidal magnetic field the magnetic field lines will move between outside and inside of the torus. This effect is described by $q = \Delta\varphi/\Delta\theta$, where $\Delta\varphi$ and $\Delta\theta$ are the changes in toroidal and poloidal angle along a field line. We realize also that the relative direction between **g** and grad n will change so that some regions have favourable and some have unfavourable curvature. In this case we may write

$$\frac{1}{R_c} = \frac{1}{R}\left(\cos\frac{2\pi z}{L} - \delta\right)$$

Since the space dependence along z is now no longer harmonic we also have to make the replacement $k_{||} \rightarrow -i\partial/\partial z$. The resulting Mathieu equation

$$\frac{\partial^2\phi}{\partial z^2} + \alpha\frac{1}{R}\left(\cos\frac{2\pi z}{L} - \delta\right)\phi = 0$$

has the eigenvalue $\delta = \alpha L^2/8\pi$ for $\alpha L^2 << 2\pi^2$. Determine the β limit in the case $\beta L^2 << 2\pi^2 aR$ when $k_y{}^2 >> k_x{}^2, \kappa \sim 1/a, \delta = a/2R$ and $L = 2\pi Rq$.
8. The approximation $E_{||} = 0$ is one of the most frequently used approximations for flute modes ($k_{||} \approx 0$). It is, however, not a good approximation for drift waves which have a slightly larger $k_{||}$. In order to see this it is necessary to consider the details of the electron dynamics. As is evident from the derivations of (2.21) and (3.29) we need only $A_{\|}$ in order to describe $\delta B_\perp$.

   (a) Show that the Boltzmann relation (3.3b) is generalized to

$$\frac{\delta n_k}{n_0} = \frac{e\phi_k}{T_e} + \frac{\omega_{\bullet e} - \omega}{k_{||}}\frac{eA_{||}}{T_e} \tag{3.70}$$

   when we include $\delta B_\perp = \nabla\times(A_{\|}\hat{z})$

$$n_e v_{||e} = -\frac{1}{e}j_{||e} \approx -\frac{1}{e}j_{||}$$

   (b) Derive another expression for $\partial n_e/\partial t$ from the electron continuity equation using i.e. neglecting the parallel ion current and express $j_{||}$ in $A_{||}$ by using Ampe'res law. Put this expression equal to (3.70) and show that

$$E_{||} = i\frac{k^2\rho^2 k_{||}{}^2 v_A{}^2}{\omega(\omega - \omega_{*e}) - k_\perp{}^2\rho_s{}^2 k_{||}{}^2 v_A{}^2}k_{||}\phi \tag{3.71}$$

(c) Eliminate $A_{||}$ and show that

$$\frac{\delta n_k}{n_0} = \frac{e\phi_k}{T_e}\left[1 + \frac{(\omega_{\bullet e} - \omega)^2}{\omega(\omega_{*e} - \omega) - k_\perp{}^2\rho_s{}^2 k_{||}{}^2 v_A{}^2}\right] \tag{3.72}$$

which turns into (3.27a) for small $k_{||}$ and into (3.3b) for large $k_{||}$.

9. Generalize (3.71) to include parallel ion motion when $v_{||_i}$ is given by $\partial v_{||i}/\partial t = (e/m_i)E_{||}$.
10. Use the tokamak parameters in Appendix I to estimate for which mode number ($k_y \approx m/a$) the FLR correction to the β limit in (3.46) exceeds 20% of the β limit for low mode numbers.
11. Include the effects of finite ion Larmor radius in the dispersion relation (3.10).
12. Show that we recover the linear dispersion relation if we impose ambipolar electron and ion fluxes from (3.57) and (3.59).

References

1. B.R. Suydam, Proc UN Int Conf on Peaceful Uses of Atomic Energy (New York: Columbia University Press), **31**, 157 (1958).
2. C. Mercier, Nucl Fus **1**, 47 (1960). L.I. Rudakov and R.Z. Sagdeev, Soviet Physics JETP {\bf 37}, 952 (1960).
3. N.A. Krall and M.N. Rosenbluth, Phys Fluids **5**, 1435 (1962).
4. A.B. Mikhailovskii and L.I. Rudakov, Soviet Physics JETP **17**, 621 (1963).
5. A.V. Timofeev, Sov Phys Tech Phys **33**, 776 (1963).
6. B.B. Kadomtsev, Plasma Turbulence, Academic Press, New York 1965.
7. N.A. Krall, Phys Fluids **9**, 820 (1966).
8. S. Moiseev, JETP Letters **4**, 55 (1966).
9. B.Lehnert, Phys. Fluids **9**, 1367 (1966).
10. BB. Kadomtsev and OP. Pogutse, Soviet Physics JETP **24**, 1172 (1967).
11. A. Simon, Phys Fluids **11**, 1186 (1968).
12. N.A. Krall in Advances in Plasma Physics (Ed A Simon and W Tomphson), Wiley, New York, Vol. **1**, p 153 (1968).
13. B.B. Kadomtsev and O.P. Pogutse in Reviews of Plasma Physics (Ed MA Leontovich) Consultants Bureau, New York, Vol **5**, p 249 (1970).
14. F.L. Hinton and C.W. Horton, Phys Fluids **14**, 116 (1971).
15. C.W. Horton and R.K. Warma, Phys Fluids **15**, 620 (1972).
16. S. Ichimaru, Basic Principles of Plasma Physics, A Statistical Approach, Benjamin, Reading 1973, Chapter 8.
17. N.A. Krall and A.W. Trivelpiece, Principles of Plasma Physics, McGraw-Hill 1973.
18. D.A. Monticello and A. Simon, Phys Fluids **17,** 791 (1974)
19. A.B. Mikhailovskii, Theory of Plasma Instabilities, Consultants Bureau, New York, Vol. **2**, 1974.
20. J.W. Connor and R.J. Hastie, Plasma Physics **17**, 97, 109 (1975).
21. G.E. Guest, C.L. Hedrik and D.B. Nelson, Phys Fluids **18**, 871 (1975).
22. A. Hasegawa, Plasma Instabilities and Nonlinear Effects, Springer 1975, Chapter 3,4.
23. A. Hasegawa and L. Chen, Phys. Rev. Lett. **35**, 370 (1975).

24. H. L. Berk, Phys Fluids **19**, 1255 (1976).
25. G. Hasselberg, A. Rogister and A. El-Nadi, Phys Fluids **20**, 982 (1977).
26. W.M. Manheimer, An introduction to Trapped Particle Instabilities in Tokamaks, ERDA Crit. Rev. Series 1977.
27. K.-I. Nichikawa, T. Hatori and Y. Terashima, Phys Fluids **21**, 1127 (1978).
28. L. Chen, J Hsu and P K Kaw, Nucl Fusion **18**, 1371 (1978).
29. W.M. Tang, Nucl. Fusion **18**, 1089 (1978).
30. W.M. Manheimer and T.M. Antonsen, Phys Fluids **22,** 957 (1979).
31. J. Weiland, H Sanuki and C S Liu, Phys Fluids **24**, 98 (1981).
32. J. Weiland, Physica Scripta **23**, 801 (1981).
33. H.U. Rahman and J Weiland, Phys Rev **A28**, 1673 (1983).
34. P.C. Liewer, Nuclear Fusion **25**, 543 (1985).
35. S.M. Kaye, Phys Fluids **28**, 2327 (1985).
36. F.W. Perkins and Y.C. Sun, Princeton Plasma Physics Laboratory Report PPPL-2216 (1985).
37. F. Romanelli, W.M. Tang and R.B. White, Nucl Fus **26**, 1515 (1986).
38. A. Rogister, G. Hasselberg, A. Kaleck, A. Boileau, H.W.H van Andel and M von Hellerman, Nucl. Fusion **26**, 797 (1986).
39. R.E. Waltz, Phys Fluids **29**, 3684 (1986).
40. W.M. Tang, Nucl. Fusion **26**, 1605 (1986).
41. W.M. Tang, G. Rewoldt and L. Chen, Phys. Fluids **29**, 3715 (1986).
42. R.R. Dominquez and R.E. Waltz, Nucl. Fusion **27**, 65 (1987).
43. A. Jarme'n, P. Andersson and J. Weiland, Nucl. Fusion **27**, 941 (1987).
44. J. Weiland and H. Nordman, Proc. Varenna-Lausanne Workshop "Theory of Fusion Plasmas", Chexbres 1988 p 451 (1988).
45. J Weiland, Comments Plasma Phys Controlled Fusion **12**, 45 (1988).
46. A.J. Wootton, M.E. Austin, R.D. Bengtson et al, Plasma Physics and Control. Fusion **30**, 1479 (1988).
47. W. Horton, B.G. Hong and W.M. Tang, Phys Fluids **31,** 2971 (1988).
48. A. Zagorodny, J. Weiland and A. Jarme'n Comments Plasma Phys Control. Fusion **17** 353 (1997).
49. A. Zagorodny and J. Weiland Ukrainian J. Physics **43**, 1402 (1998).

Chapter 4
Kinetic Description of Low Frequency Modes in Inhomogeneous Plasma

4.1 Integration Along Unperturbed Orbits

In the previous chapter we derived simple dispersion relations for some of the most dangerous low frequency instabilities using a fluid description. We will now show how this can be done by kinetic theory, [1–19], from the Vlasov equation in a simple slab geometry. We will start by using the method of integration along unperturbed orbits [1–5], which gives the most general result, i.e. including also modes with $\omega \geq \Omega_c$, full finite Larmor radius effects and wave particle resonances. We will, however, restrict attention to modes with $\omega << \Omega_c$. We will show how wave-particle resonances may impede the free electron motion along the field lines, thus causing drift instability and how the lowest order finite Larmor radius (FLR) effect agrees with that obtained from the stress tensor in Chap. 2. After the more general treatment we will show how the wave-particle resonances can be described by a simpler drift-kinetic equation that does not contain FLR effects and how the lowest order FLR effect can be obtained by a simple orbit averaging.

As pointed out previously the parallel motion of electrons may be affected by wave-particle resonances and also by inductance. We are now going to give a kinetic description of these phenomena. The first problem that arises is to determine the unperturbed distribution function $f_0(v,x)$, where we choose the inhomogeneity to be the x direction. For generality we include in our description also a gravitational force acting in the x direction. Inhomogeneities in the externally produced magnetic field may be included in this gravitational force, as well as a centrifugal force due to the toroidicity. The unperturbed distribution function $f_0(v,x)$ may be written as a function of the constants of motion

$$W = \frac{1}{2}mv^2 - mgx; P_y = m(v_y + \Omega_c x) \qquad \text{and} \quad P_{||} = mv_{||} \tag{4.1}$$

J. Weiland, *Stability and Transport in Magnetic Confinement Systems*, Springer Series on Atomic, Optical, and Plasma Physics 71,
DOI 10.1007/978-1-4614-3743-7_4, © Springer Science+Business Media New York 2012

The constant P_y can easily be derived from the equation of motion in the form

$$\frac{d\mathbf{v}}{dt} = \Omega_c(\mathbf{v} \times \widehat{\mathbf{z}})$$

$$\frac{dv_y}{dt} = -\Omega_c v_x = -\Omega_c \frac{dx}{dt}$$

Here we assumed $\Omega_c = qB/m$ to be homogeneous. A *Maxwellian distribution function* corresponding to an exponential density gradient may be written:

$$f_0(\mathbf{v},x) = n_0\left(\frac{m}{2\pi T}\right)^{3/2} \exp\left[-\alpha'\left(x + \frac{v_y}{\Omega_c}\right)\right] \exp\left[-\frac{mv^2/2 - mgx}{T}\right] \tag{4.2}$$

It is important to note that Ω_c here contains the sign of the charge. For a plasma with perpendicular and parallel temperature gradients we may replace α' by $\alpha' + \delta_\perp v_\perp{}^2 + \delta_{||} v_{||}{}^2$. We will, however, for simplicity neglect temperature gradients. For a weak inhomogeneity, i.e. small α' we may expand the first exponential in (4.2). To first order in α' and g we then obtain the *zero order drift velocity*.

$$\mathbf{v_d} = \frac{1}{n_0}\int \mathbf{v} f_0(\mathbf{v},x)d\mathbf{v} = -\frac{\alpha' T}{m\Omega_c}\widehat{\mathbf{y}} = -\left(\frac{\kappa T}{m\Omega_c} + \frac{g}{\Omega_c}\right)\widehat{\mathbf{y}} \tag{4.3}$$

where $\kappa = -(1/n_0)dn_0/dx$. We note that in this expression we have to take Ω_c negative for electrons. For simplicity we will in the following continue to use the previous assumption that $k_x = 0$. We may then write

$$\mathbf{E}(\mathbf{r},t) = \mathbf{E_k} e^{i(k_y y + k_{||} z - \omega t)} + c.c.$$

$$f(\mathbf{r},\mathbf{v},t) = f_{\mathbf{k}}(\mathbf{v}) e^{i(k_y y + k_{||} z - \omega t)} + c.c.$$

The linearized Vlasovequation may then be written:

$$\left[\frac{\partial}{\partial t} + \mathbf{v}\cdot\frac{\partial}{\partial \mathbf{r}} + \left[\Omega_c(\mathbf{v}\times\widehat{\mathbf{z}}) + g\widehat{\mathbf{x}}\right]\cdot\frac{\partial}{\partial \mathbf{v}}\right] f_k(\mathbf{v}) e^{i(k_y y + k_{||} z - \omega t)}$$
$$= \frac{q}{m}(\mathbf{E_k} + \mathbf{v}\times\mathbf{B_k})\cdot\frac{\partial f_0}{\partial \mathbf{v}} e^{i(k_y y + k_{||} z - \omega t)} \tag{4.4}$$

The left hand side of (4.4) is here the total derivative along a particle orbit. In a linear approximation we use the unperturbed orbit

$$\mathbf{v}' = \mathbf{v}(t') = \widetilde{\Lambda}(t' - t)[\mathbf{v}(t) - \mathbf{v_g}] + \mathbf{v_g}$$

$$\mathbf{r}' = \mathbf{r}(t') = \widetilde{\Gamma}(t' - t)[\mathbf{v}(t) - \mathbf{v_g}] + \mathbf{v_g}(t' - t)$$

Where $\mathbf{v_g} = -(g/\Omega_c)\,\hat{\mathbf{y}}$ and

$$\tilde{\Gamma}(t) = \begin{vmatrix} \sin\Omega_c t & 1-\cos\Omega_c t & 0 \\ \cos\Omega_c t - 1 & \sin\Omega_c t & 0 \\ 0 & 0 & \Omega_c t \end{vmatrix}$$

Where

$$\tilde{\Lambda} = \frac{1}{\Omega_c}\frac{d\tilde{\Gamma}}{dt} \tag{4.5}$$

Integration along this orbit yields

$$f_k(\mathbf{v}) = \frac{q}{m}\int_0^\infty d\tau(\mathbf{E_k} + \mathbf{v}\times\mathbf{B_k})\cdot\frac{\partial f_0}{\partial \mathbf{v}}e^{-i\alpha(\tau)} \tag{4.6}$$

Where

$$\alpha(\tau) = \int_0^\tau [\mathbf{k}\cdot\mathbf{v} - \omega]d\tau = k_y\left[\frac{v_x}{\Omega_c}(1-\cos\Omega_c\tau) + \frac{v_y - v_g}{\Omega_c}\sin(\Omega_c\tau) + v_g\tau\right] - \omega\tau$$

And $\tau = t - t'$

We will here consider the region $\beta << 1(\beta = 2\mu_0 nT/B^2)$. In this region $k_\perp v_A >> k_\perp v_{Di}$ and we may disregard the compressional (magnetosonic) wave. The only electromagnetic effect is then due to the bending of the magnetic field lines and we have

$$\delta\mathbf{B}_{||} = (\nabla\times\mathbf{A})_{||} = \frac{\partial A_x}{\partial y} - \frac{\partial A_y}{\partial x} = 0$$

for the perturbed field. This means that we can derive the perpendicular part of **A** from a potential. Such a potential has sometimes been introduced. Here, we will instead continue to use $A_{||}$ which is a good low β approximation. Thus we write:

$$\mathbf{E} = -\nabla\phi - \frac{\partial\mathbf{A}}{\partial t} = -\nabla\phi - \hat{\mathbf{z}}\frac{\partial A_{||}}{\partial t} \tag{4.7}$$

and thus

$$\frac{\partial\mathbf{B}}{\partial t} = -(\hat{\mathbf{z}}\times\nabla_\perp)\frac{\partial A_{||}}{\partial t}$$

Now

$$\frac{\partial f_0}{\partial\mathbf{v}} = -\left(\frac{\alpha'}{\Omega_c}\hat{\mathbf{y}} + \frac{m}{T}\mathbf{v}\right)f_0(x,\mathbf{v}) \tag{4.8}$$

Since $k_x = 0$ we may rewrite (4.6) as

$$\begin{aligned} f_k(\mathbf{v}) = &-\frac{q}{m}\int_0^\infty i\left[\frac{m}{T}\left((k_y \mathrm{v_y}' + k_{||}\mathrm{v}_{||}')\phi + \omega \mathrm{v}_{||}' A_{||}\right) + \frac{\alpha'}{\Omega_c}k_y\phi\right]\cdot f_0 e^{-i\alpha(\tau)} \\ &- i\frac{q}{m}\int_0^\infty \left[\frac{\alpha'}{\Omega_c}k_y(\mathbf{v}\times\widehat{\mathbf{x}})\cdot\widehat{\mathbf{y}}A_{||}\right]\cdot f_0 e^{-i\alpha(\tau)} \end{aligned} \tag{4.9}$$

Since

$$\frac{d}{d\tau}e^{-i\alpha(\tau)} = -i\left[k_y \mathrm{v_y}' + k_{||}\mathrm{v}_{||}' - \omega\right]e^{-i\alpha(\tau)} \tag{4.10}$$

It is convenient to rewrite (4.9) as

$$\begin{aligned} f_k(\mathbf{v}) = &-\frac{q\phi}{T}\int_0^\infty i\left[k_y \mathrm{v_y}' + k_{||}\mathrm{v}_{||}' - \omega\cdot\right]\cdot f_0 e^{-i\alpha(\tau)}d\tau \\ &-\frac{q\phi}{T}\int_0^\infty i[(\omega - \omega_{Ds}\cdot)]\cdot f_0{}^{-i\alpha(\tau)}d\tau \\ &-\frac{qA_{||}}{T}\int_0^\infty i(\omega - \omega_{Ds})\mathrm{v}_{||}'\cdot f_0{}^{-i\alpha(\tau)}d\tau \end{aligned} \tag{4.11}$$

where

$$\omega_{Ds} = -k_y\frac{\alpha' T}{m\Omega_c} = \omega_* - k_y \mathrm{v_g} \qquad \mathrm{v_g} = \frac{g}{\Omega_c}$$

With the help of (4.10) we can immediately integrate the first integral of (4.11). We note that the limit $\tau\to\infty$ corresponds to a contribution from the perturbation at $t'\to-\infty$. We take this to be zero. Observing that the unperturbed distribution function is invariant along an unperturbed orbit we then find

$$\begin{aligned} f_k(\mathbf{v}) = &-\frac{q}{T}f_0(x,\mathbf{v}) \\ &\times\left\{\phi + i\left[(\omega - \omega_{Ds})\phi + \omega\cdot \mathrm{v}_{||}A_{||}\left(1 - \frac{\omega_{Ds}}{\omega}\right)\right]\int_0^\infty e^{-i\alpha(\tau)}\cdot d\tau\right\} \end{aligned} \tag{4.12}$$

In order to evaluate the integral we now make use of the expansion

$$\begin{aligned} e^{-i\alpha(\tau)} = &\sum_{n=-\infty}^{\infty}\sum_{n'=-\infty}^{\infty} J_n(\xi)J_{n'}(\xi) \\ &\times \exp\left\{i\left[n\left(\Omega_c\tau + \theta + \frac{\pi}{2}\right) - n'\theta + k_{||}\mathrm{v}_{||}\tau - \tilde{\omega}\tau\right]\right\} \end{aligned} \tag{4.13}$$

Where

$$\tilde{\omega} = \omega + k_y \mathrm{v_g} \qquad \xi = k_y \mathrm{v}_\perp / \Omega_c$$

We then obtain

$$f_k(\mathbf{v}) = -\frac{q}{T} f_0(x, \mathbf{v}) \Big\{ \phi + i \Big[(\omega - \omega_{Ds}) \phi + \omega \cdot \mathrm{v}_{||} A_{||} \Big(1 - \frac{\omega_{Ds}}{\omega} \Big) \Big] \times \sum_{n,n'} \frac{J_n(\xi) J_{n'}(\xi) e^{-i(n-n')\theta}}{n\Omega_c + k_{||}\mathrm{v}_{||} - \tilde{\omega}} \Big\} \tag{4.14}$$

We can now obtain the dispersion relation from the Maxwells equations

$$\nabla \cdot \mathbf{E} = \frac{\rho}{\varepsilon_0} \tag{4.15}$$

$$(\nabla \times \mathbf{B})_{||} = \mu_0 j_{||} \tag{4.16}$$

By using the formula

$$\int_0^\infty e^{-a^2 x^2} x J_n(px) J_n(qx) dx = \frac{1}{2a^2} \exp\left[-\frac{p^2 + q^2}{4a^2} \right] I_n\left(\frac{pq}{2a^2} \right)$$

where I_n is a modified Bessel function, it is possible to show that

$$\sum_{n=-\infty}^{\infty} \sum_{n'=-\infty}^{\infty} \int \frac{f_0 J_n(\xi) J_{n'}(\xi) e^{-i(n-n')\theta} d\mathrm{v}}{n\Omega_c + k_{||}\mathrm{v}_{||} - \tilde{\omega}} = \sum_n \frac{\Lambda_n(\varsigma)}{\tilde{\omega} - n\Omega_c} \left[W\left(\frac{\tilde{\omega} - n\Omega_c}{|k_{||}|(T/m)^{1/2}} \right) - 1 \right]$$

where $\Lambda_\mathrm{n}(\varsigma) = I_n(\varsigma) \mathrm{e}^{-\varsigma}$. W(z) is the plasma dispersion function

$$W(z) = (2\pi)^{1/2} \int_{-\infty}^{\infty} \frac{x}{x - z} e^{-x^2/2} dx$$

and $\varsigma = \mathrm{k}_\perp{}^2 \mathrm{T} / \mathrm{m}\Omega_\mathrm{c}{}^2$

We thus obtain

$$n_k = \int f_k(\mathbf{v}) d\mathrm{v} = -\frac{q n_0}{T} \left\{ \phi + (\omega - \omega_{Ds}) \phi \times \sum_n \frac{\Lambda_n(\varsigma)}{\tilde{\omega} - n\Omega_c} \left[W\left(\frac{\tilde{\omega} - n\Omega_c}{|k_{||}|(T/m)^{1/2}} \right) - 1 \right] \right\} - \frac{q n_0}{T} \omega \mathrm{v}_{||} A_{||} \left(1 - \frac{\omega_{Ds}}{\omega} \right) \sum_{n,n'} \Lambda_n(\varsigma) W\left(\frac{\tilde{\omega} - n\Omega_c}{|k_{||}|(T/m)^{1/2}} \right) \tag{4.17}$$

Similarly we obtain

$$j_{\|k} = q\int v_z f_k(\mathbf{v})dv = -\frac{q^2 n_0}{Tk_\|}(\omega - \omega_{Ds})\phi \times \sum_{n,n'}\Lambda_n(\varsigma)W\left(\frac{\tilde{\omega} - n\Omega_c}{|k_\||(T/m)^{1/2}}\right)$$
$$-\frac{q^2 n_0}{T}\frac{\omega}{k_\|^2}\cdot A_\|\left(1 - \frac{\omega_{Ds}}{\omega}\right)\sum_{n,n'}\Lambda_n(\zeta)(\tilde{\omega} - n\Omega_c)W\left(\frac{\tilde{\omega} - n\Omega_c}{|k_\||(T/m)^{1/2}}\right) \tag{4.18}$$

By using the density and current perturbations (4.17) and (4.18) we obtain the dispersion relation (4.16). In the following we will mainly consider the frequency range $\omega << n\Omega_c$. In this region we need to include only the term $n = 0$ in the summations of (4.17) and (4.18). For the last term in (4.18) this is due to the adiabatic expansion (4.21). We further note that $\varsigma = k_\perp T/m\Omega_c^2 = k_\perp^2\rho^2/2$ expresses the ratio between the Larmor radius ρ and the perpendicular wavelength. When ς is small we may use the expansion

$$\Lambda_0(\varsigma) \approx 1 - \varsigma \quad (\varsigma << 1) \tag{4.19}$$

The deviation of $\Lambda_0(\varsigma)$ from 1 will in the following be referred to as a finite Larmor radius effect. It is due to the fact that a particle that gyrates in a Larmor orbit on the average experiences an electric field that is different from the field at the center of the orbit. We will always assume that $\Lambda_0(\varsigma) = 1$ for electrons, while we will make different approximations for the ions.

We shall assume that (3.6) is valid also when we replace ω by the shifted frequency $\omega - k_y v_g$. The plasma dispersion function is usually expanded in the adiabatic $\omega >> k_\| v_{th}$ and isothermal $\omega << k_\| v_{th}$ limits. These expansions are:

$$W(z) = i\left(\frac{\pi}{2}\right)^{1/2} z e^{-z^2/2} + 1 - z^2 + z^4/4 \ldots\ldots |z| << 1 \qquad \text{(isothermal)} \tag{4.20}$$

and

$$W(z) = i\left(\frac{\pi}{2}\right)^{1/2} z e^{-z^2/2} - \frac{1}{z^2} - \frac{3}{z^4} \ldots\ldots |z| >> 1 \qquad \text{(adiabatic)} \tag{4.21}$$

Now ignoring $k_y v_{ge}$ (electron gravity) and using the isothermal expansion for electrons and the adiabatic expansion for ions (compare Eq. 3.1) we can write the particle densities

$$\frac{n_{ke}}{n_0} = \frac{e}{T_e}\left\{\begin{array}{l} \phi + i\left(\frac{\pi}{2}\right)^{1/2}\dfrac{\omega - \omega_{*e}}{|k_\||(T_e/m_e)^{1/2}} e^{-\omega^2/\left(k_\|^2 v_{the}^2\right)}\phi \\ + \dfrac{\omega}{k_\|}A_\|\left(1 - \dfrac{\omega_{*e}}{\omega}\right)\left[1 + i\left(\dfrac{\pi}{2}\right)^{1/2}\dfrac{\omega}{|k_\||(T_e/m_e)^{1/2}} e^{-\omega^2/\left(k_\|^2 v_{the}^2\right)}\right] \end{array}\right\} \tag{4.22}$$

$$\frac{n_{ki}}{n_0} = -\frac{e}{T_e}$$

$$\times \left\{ \begin{aligned} &\phi - (\omega - \omega_{Dsi})\frac{\Lambda_0(\varsigma_i)}{\tilde{\omega}}\left[1 + \frac{k_{\|}{}^2 T_i}{m_i \tilde{\omega}^2} - i\left(\frac{\pi}{2}\right)^{1/2} \frac{\tilde{\omega}}{|k_{\|}|(T_i/m_i)^{1/2}} e^{-\omega^2/\left(k_{\|}{}^2 \mathrm{v_{the}}^2\right)}\right] \\ &+ \frac{\omega}{k_{\|}} A_{\|}\left(1 - \frac{\omega_{Dsi}}{\omega}\right)\Lambda_0(\varsigma_i)\left[i\left(\frac{\pi}{2}\right)^{1/2} \frac{\tilde{\omega}}{|k_{\|}|(T_i/m_i)^{1/2}} e^{-\omega^2/\left(k_{\|}{}^2 \mathrm{v_{thi}}^2\right)} - \frac{k_{\|}{}^2 T_i}{m_i \tilde{\omega}^2}\right] \end{aligned} \right\}$$

(4.23)

And the parallel electron current

$$j_{\|k} = -\frac{e^2 n_0}{T_e k_{\|}}\left(\phi + \frac{\omega}{k_{\|}} A_{\|}\right)(\omega - \omega_{*e}) \times \left[1 + i\left(\frac{\pi}{2}\right)^{1/2} \frac{\omega}{|k_{\|}|(T_e/m_e)^{1/2}} e^{-\omega^2/\left(k_{\|}{}^2 \mathrm{v_{the}}^2\right)}\right] \tag{4.24}$$

In (4.22) we recognise the Bolzmann distribution in the first term of the right hand side. The second term represents the phase shift due to wave-particle resonance and the third term a correction due to induction.

4.2 Universal Instability

The electrostatic limit is easily obtained from the above equations by putting $A_{\|} = 0$. The dispersion relation obtained from (4.15) can be written in the form

$$\varepsilon(k_y, k_{\|}, \omega) = 0 \tag{4.25}$$

where

$$\varepsilon = \frac{\phi_{ind}}{\phi_{ext}}$$

Here φ_{ind} is the induced potential caused by an external potential φ_{ext}. In the region (3.1) we obtain

$$\begin{aligned} \varepsilon(k_y, k_{\|}, \omega) = 1 &+ \frac{k_{de}{}^2}{k_y{}^2}\left[1 + i\left(\frac{\pi}{2}\right)^{1/2} \frac{\omega - \omega_{*e}}{|k_{\|}|(T_e/m_e)^{1/2}} e^{-\omega^2/\left(k_{\|}{}^2 \mathrm{v_{the}}^2\right)}\right] \\ &+ \frac{k_{di}{}^2}{k_y{}^2}\left\{1 - \frac{\omega - \omega_{Dsi}}{\tilde{\omega}}\left[1 + \frac{k_{\|}{}^2 T_i}{m_i \tilde{\omega}^2} - i\left(\frac{\pi}{2}\right)^{1/2} \frac{\tilde{\omega}}{|k_{\|}|(T_i/m_i)^{1/2}} e^{-\omega^2/\left(k_{\|}{}^2 \mathrm{v_{thi}}^2\right)}\right]\Lambda_0(\varsigma_i)\right\} \end{aligned} \tag{4.26}$$

In the limit $k\lambda_d << 1$ the 1 is negligible in (4.26) and the corresponding dispersion relation (4.25) can be obtained by using quasineutrality. Assuming the wave particle interaction to be weak we can treat imaginary parts of ε as perturbations. Thus writing $\varepsilon = \varepsilon_R + i\varepsilon_I$ we can solve (4.26) in the usual way as $\omega = \omega_r + i\gamma$ where

$$\varepsilon_R(k_y, k_{||}, \omega_r) = 0 \tag{4.27}$$

and

$$\gamma = -\frac{\varepsilon_I(k_y, k_{||}, \omega_r)}{\partial\varepsilon_R/\partial\omega} \tag{4.28}$$

From (4.27) we obtain the relation

$$1 + \frac{T_e}{T_i} - \frac{\omega - \omega_{*i}}{\tilde{\omega}}\left[\frac{T_e}{T_i} + \frac{k_{||}{}^2 c_s{}^2}{\tilde{\omega}^2}\right]\Lambda_0(\varsigma_i) = 0 \tag{4.29}$$

For small $k_{||}$ and dropping the gravity we obtain the solution

$$\omega = \frac{\omega_{*e}\Lambda_0(\varsigma_i)}{1 + (T_e/T_i)[1 - \Lambda_0(\varsigma_i)]} \tag{4.30}$$

where we used the relation

$$\omega_{*i} = -\frac{T_i}{T_e}\omega_{*e}$$

By expanding $\Lambda_0(\varsigma)$ according to (4.19) and introducing $\varsigma_i = k_y{}^2\rho_i{}^2/2$, we obtain the solution

$$\omega = \omega_{*e}\left[1 - \frac{1}{2}k_y{}^2\rho_i{}^2\left(1 + \frac{T_e}{T_i}\right)\right] \tag{4.31}$$

Since $\rho^2 = (T_e/2T_i)\rho_i{}^2$, we recognise the last term to be due to inertia (polarisation drift) by comparison with (3.10). The result (4.31) shows that ion FLR effects can be added in a simple way for drift waves. The similarity between ion polarization drift and ion FLR effects can be seen from that the polarization drift comes from the variation in time of the electric field along a gyroorbit while the FLR effect comes from the variation in space of the electric field along the gyroorbit. The rotating particle just sees a varying electric field along the orbit in both cases. Clearly the ion inertia dominates the Larmor radius effect (FLR) when $T_e >> T_i$. From (4.28) we find

$$\gamma = -\left(\frac{\pi}{2}\right)^{1/2} \frac{\omega_r{}^2}{\omega_{*e}\Lambda_0(\varsigma_i)}$$
$$\times \left[\frac{\omega_r - \omega_{*e}}{|k_{||}|(T_e/m_e)^{1/2}} e^{-\omega^2/\left(k_{||}{}^2 \mathrm{v_{the}}{}^2\right)} + \frac{T_e}{T_i}\left(\frac{\pi}{2}\right)^{1/2} \frac{\omega_r - \omega_{*i}}{|k_{||}|(T_i/m_i)^{1/2}} e^{-\omega^2/\left(k_{||}{}^2 \mathrm{v_{thi}}{}^2\right)}\right] \tag{4.32a}$$

We notice that the situation concerning stability is similar to that for the influence of ion-electron collisions. Due to FLR effects we note from (4.31) that $\omega<\omega_{*e}$. According to (4.32) this means that the interaction between the wave and the electrons is destabilising. Sinceω_{*i}, <0 for κ>0 we find that the ions will cause damping. Due to the exponential factor, however, this term will usually be small in the region (3.1). The collisionless instability described by (4.30) and (4.32) is usually referred to as the *universal instability* since it was for a long time considered to be unavoidable in a finite size plasma. We note, however, that the Landau-damping term will become important in a short device where $k_{||}$ has to take rather large values. Moreover, in a device with magnetic shear, $k_{||}$ can take small values only locally and damping is obtained by convection into regions with larger $k_{||}$. Another situation when growth and damping can alternate is when we have a nonlinear frequency shift. We can understand that from the previous discussion but this becomes even more easy to see if we simplify (4.32) as:

$$\gamma = \left(\frac{\pi}{2}\right)^{1/2} \omega_{*e} \frac{\omega - \omega_{*e}}{k_{||}\mathrm{v}_{te}} e^{-\omega^2/\left(k_{||}v_{te}\right)} \tag{4.32b}$$

where we ignored FLR effects and ion Landau damping. This will be discussed further in connection with the fluid closure. Finally we note that since $\omega - \omega_{\mathrm{Dsi}} = \omega + \mathrm{k_y v_{gi}} - \omega_{*\mathrm{i}}$,the only effect of gravity on the dispersion relation (4.29) is a shift of the real part of ω equivalent to a change of frame. Thus in a frame moving with a velocity $\mathbf{v_g} = -(\mathrm{g}/\Omega_{\mathrm{ci}})\hat{\mathbf{y}}$, the dispersion relation will take the same form as in the laboratory frame when the gravity is absent. This is, however, only true as long as we may drop the frequency dependent terms in the electron part of the dispersion relation, i.e. as long as the electrons are able to maintain a Boltzmann distribution. For very small $k_{||}$ this is no longer possible and we obtain a reactive instability called Rayleigh-Taylor or Interchange instability.

4.3 Interchange Instability

In the limit $k_{||} = 0$ we have $W(z_e) = W(z_i) = 0$. The electrostatic dispersion relation then takes the form

$$\frac{1}{T_e}\frac{\omega_{*e}}{\omega} = -\frac{1}{T_i}\left[1 - \frac{\tilde{\omega} - \omega_{*i}}{\tilde{\omega}}\Lambda_0(\varsigma_i)\right] \tag{4.33}$$

Introducing

$$\tilde{\omega} = \omega + k_y \frac{g}{\Omega_{ci}} = \omega - k_y \mathrm{v}_{gi}$$

where $\mathrm{v}_{gi} = -g/\Omega_{ci}$ is the gravitational drift and $\omega_{*i} = -(T_i/T_e)\omega_{*e}$, we may rewrite (4.33) as

$$\omega\omega_{*e}(1 - \Lambda_0) - \omega_{*e} k_y \mathrm{v}_{gi} + \frac{T_e}{T_i}(1 - \Lambda_0)\omega(\omega - k_y \mathrm{v}_{gi}) = 0 \tag{4.34}$$

Introducing

$$\omega_{*e} = k_y \frac{\kappa T_e}{e B_0}$$

in the constant term we finally arrive at the dispersion relation

$$\omega\left[\omega - k_y(\mathrm{v}_{gi} + \mathrm{v}_{*i})\right] = -\kappa g \frac{k^2 \rho_i^2}{1 - \Lambda_0} \tag{4.35}$$

This dispersion relation predicts instability when

$$\kappa g \frac{k^2 \rho_i}{1 - \Lambda_0} > \frac{1}{4} k_y{}^2 (\mathrm{v}_{gi} + \mathrm{v}_{*i})^2 \tag{4.36}$$

This instability is due to the charge separation created by a density perturbation when the electron and ion guiding centre drifts are different. Since $k_{||} = 0$ the electrons cannot shield the charge separation by moving along B_0. This is the Rayleigh-Taylor or Interchange Instability. Clearly a condition for a possibility to fulfill (4.36) is that $\kappa\, g>0$, i.e. the gravity and density gradient have opposite direction as shown in Fig. 3.9.

In the unstable case the more dense parts tend to change place (interchange) with the less dense parts thus causing a convective diffusion. When grad n and **g** have the same direction a perturbation is counteracted and this results in oscillations. The instability is analogous to that of a heavy fluid resting on top of a light fluid. In a toroidal machine the centrifugal force due to the field line curvature may give rise to interchange instability in regions of unfavourable curvature ($\kappa g>0$). We note that also when the different drift velocities of electrons and ions are caused by a gravitational force, the finite Larmor radius effect is stabilizing, contrary to the situation for drift waves.

4.4 Drift Alfvén Waves and β Limitation

We will now consider the electromagnetic case in the region (3.1). We write the equations in a frame where the background guiding centre drift of the electrons is zero. In this frame the ion background drift will be equal to the difference between the ion and electron drifts in the laboratory frame. We will assume that we can neglect the imaginary part of W(z) both for ions and electrons. We then have from (4.20) and (4.21)

$$\frac{n_e}{n_0} = \frac{e}{T_e}\left[\phi + \frac{\omega - \omega_{*e}}{k_{||}} A_{||}\right] \tag{4.37}$$

$$\frac{n_i}{n_0} = -\frac{e}{T_i}\left[1 - (\omega - \omega_{Dsi})\frac{\Lambda_0}{\tilde{\omega}}\right]\phi \tag{4.38}$$

$$j_{||} = \frac{e^2 n_0}{T_e k_{||}}(\omega_{*e} - \omega)\left(\phi + \frac{\omega}{k_{||}} A_{||}\right) \tag{4.39}$$

From the induction law, (4.16) we obtain

$$j_{||} = -\frac{k_\perp{}^2}{\mu_0} A_{||} \tag{4.40}$$

Neglecting parallel ion motion we now obtain from (4.39) and (4.40)

$$A_{||} = \frac{k_{||}(\omega_{*e} - \omega)}{\omega(\omega_{*e} - \omega) + k_\perp{}^2 \rho_s{}^2 k_{||}{}^2 \mathrm{v_A}{}^2}\phi \tag{4.41}$$

Equation 4.41 gives the relation between the parallel and the perpendicular electric fields. It involves the electron dynamics and can easily be obtained from the fluid equations. Inserting now (4.41) into (4.37) we obtain the electron response in terms of φ.

$$\frac{n_e}{n_0} = \frac{e\phi}{T_e}\left[\frac{\omega_{*e}}{\omega} + \left(1 - \frac{\omega_{*e}}{\omega}\right)\frac{k_\perp{}^2 \rho_s{}^2 k_{||}{}^2 \mathrm{v_A}{}^2}{\omega(\omega_{*e} - \omega) + k_\perp{}^2 \rho_s{}^2 k_{||}{}^2 \mathrm{v_A}{}^2}\right] \tag{4.42}$$

Equation 4.42 which is equivalent to (3.72) shows that for large $k_{||}\mathrm{v_A}$ the electron response approaches a Boltzmann distribution while it in the opposite case is of the flute mode type, i.e. proportional to ω_{*e}/ω. Introducing $\omega - \omega_{Dsi} = \omega - k_y v_{gi} - \omega_{*i}$ and $\omega_{*i} = -(T_i/T_e)\omega_{*e}$ we find from (4.38)

$$\frac{n_i}{n_0} = \frac{e\phi}{T_e}\left[\frac{\omega_{*e}}{\tilde{\omega}}\Lambda_0 - \frac{T_e}{T_i}(1 - \Lambda_0)\right] \tag{4.43}$$

We may now obtain a dispersion relation for drift Alfve'n waves from (4.42) and (4.43) assuming quasineutrality.

$$\left(\omega_{*e} + \frac{T_e}{T_i}\tilde{\omega}\right)(\Lambda_0 - 1) + \omega_{*e}\frac{k_y \mathrm{v_g}}{\omega} = (\omega - \omega_{*e})\frac{k_\perp{}^2\rho_s{}^2 k_{||}{}^2 \mathrm{v_A}^2}{\omega(\omega_{*e} - \omega) + k_\perp{}^2\rho_s{}^2 k_{||}{}^2 \mathrm{v_A}^2}$$

We now multiply by the denominator in the right side, divide by $(1 - \Lambda_0)$ and multiply by T_i/T_e. Observing now that $v_g = v_{gi} - v_{ge}$ and $v_{gj} = g_j/\Omega_{cj}$ we may write

$$\omega_{*e}k_y \mathrm{v_g} = -k_\perp{}^2\rho_s{}^2\kappa g_i\left(1 + \frac{m_e g_e}{m_i g_i}\right) \tag{4.44}$$

and obtain the equation

$$\omega[\omega - k_y(\mathrm{v_{gi}} + \mathrm{v_{*i}})] - k_{||}{}^2\mathrm{v_A}^2\frac{1}{2}\frac{k_\perp{}^2\rho_i{}^2}{1 - \Lambda_0}\left(1 - \frac{k_y \mathrm{v_g}}{\omega - \omega_{*e}}\right) + \kappa g_i\left(1 + \frac{m_e g_e}{m_i g_i}\right)\frac{1}{2}\frac{k_\perp{}^2\rho_i{}^2}{1 - \Lambda_0} =$$
$$= k_\perp{}^2\rho_s{}^2 k_{||}{}^2\mathrm{v_A}^2\frac{\omega - k_y(\mathrm{v_g} + \mathrm{v_{*i}})}{(\omega - \omega_{*e})} \tag{4.45}$$

Equation 4.45 is the dispersion relation for drift Alfve'n waves including a gravitational force and full finite Larmor radius effects. We notice that (4.45) is identical to (4.35) in the limit $k_{||} = 0$. We are, however, not allowed to take this limit of (4.45) since it would correspond to an expansion for $\omega/k_{||} >> v_{the}$. The reason why it gives the correct result is that the electron density distributions become the same for the two cases as seen from (4.42) and from (4.17). We also notice that the flute mode response is obtained in the limit $E_{||} \sim k_{||}\varphi - \omega A_{||} = 0$. In this case the induction force prevents the electrons from cancelling space charge by moving along B_0 and this makes the interchange mode solution possible.

Clearly the Alfvén frequency $k_{||}v_A$ has a stabilizing influence on the interchange instability. This can be seen as a result of the bending of the frozen in magnetic field lines which counteracts the interchange of fluid elements. The balance between these forces leads to a β limit for stability discussed in Chap. 2. The drift terms are also stabilizing. The $k_y v_{*i}$ term is due to reduction of the convective E × B drift velocity of ions when averaged over a Larmor orbit and leads to a stabilizing charge separation effect, compare Sect. 3.3.7. The most unstable situation will obviously occur for small k_y. We also note the term due to v_g in the Alfvén term. This term is often considered to be small and, in fact, should be small in the present treatment

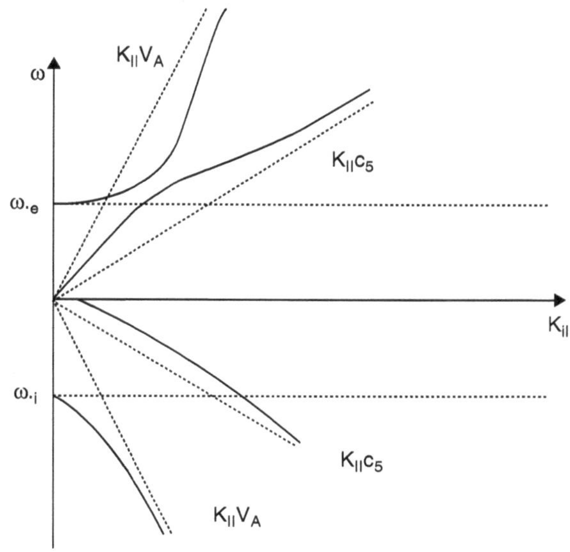

Fig. 4.1 Dispersion diagram for electromagnetic drift waves

since we represented curvature by a gravity force. However, this term is also present if we use real curvature and gradient B effects. It may then be important.

If we expand (4.45) for small ion temperature keeping only the lowest order Larmor radius effects and also neglect terms of order $k_y v_g/\omega$ we obtain the dispersion relation

$$\omega(\omega - \omega_{*i}) - k_{||}^2 v_A^2 + \kappa g_i\left(1 + \frac{m_e g_e}{m_i g_i}\right) = k_\perp^2 \rho_s^2 k_{||}^2 v_A^2 \frac{\omega - \omega_{*i}}{\omega - \omega_{*e}} \qquad (4.46)$$

Equation 4.46 agrees with (3.44) for small electron temperature if the gravity is expressed as a centrifugal acceleration and thus verifies the lowest order finite Larmor radius effect as obtained from the stress tensor. The right hand side of (4.46) is due to the parallel electric field and provides a coupling to the electrostatic drift wave branch. In studying this coupling we will for simplicity neglect the gravity.

Assuming $k_y^2 \rho_i^2 << 1$ we then realise that (4.46) splits into two branches the electric drift wave branch with $\omega = \omega_{*e}$ and the electromagnetic drift wave branch or drift Alfve'n branch.

If we include the term proportional to $k_{||}^2 T_i/m_i\omega^2$ in (4.23), (4.46) generalizes to

$$\left[\omega(\omega - \omega_{*e}) - k_{||}^2 c_s^2\right]\left[\omega(\omega - \omega_{*i}) - k_{||}^2 v_A\right] = k_\perp^2 \rho_s^2 k_{||}^2 v_A^2 \omega(\omega - \omega_{*i}) \qquad (4.47)$$

This dispersion relation shows the coupling between the drift acoustic and drift Alfvén branches. It has four branches as shown by Fig. 4.1.

We notice from this figure where $v_A > c_s$ that instead of an intersection of the branches corresponding to $k_y^2 \rho_s^2 = 0$ the branches change their identity and we obtain a region of strong coupling. The condition for the existence of this region is clearly $v_A > c_s$. On the other hand, in order to remain in the region (3.1) of drift waves we must have $v_A < v_{the}$. This condition is equivalent to $\beta > m_e/m_i$, which is the limit of β where the electromagnetic effects have to be included.

4.5 Landau Damping

If we now return to (4.46), neglect the right hand side but include the electron and ion Landau damping effects from (4.22) and (4.23) to leading order, we obtain the dispersion

$$\omega(\omega - \omega_{*i}) - k_{||}^2 v_A^2 + D = i\omega\Gamma \tag{4.48}$$

where

$$D = \kappa g_i \left(1 + \frac{m_e g_e}{m_i g_i}\right)$$

and

$$\Gamma = \left(\frac{\pi}{2}\right)^{1/2} k_{||} v_A \frac{v_A}{c_s} \left[\left(\frac{T_e}{T_i}\right)^{3/2} \frac{\omega - \omega_{*i}}{\omega - \omega_{*e}} e^{-\tilde{\omega}^2/(k_{||}^2 v_{ti}^2)} + \sqrt{m_e/m_i}\right]$$

Assuming now that $\omega = \omega_r + i\gamma$ and $\gamma << \omega_r$ we obtain by separation

$$\omega_r = \frac{1}{2}\omega_{*i} \pm \sqrt{\frac{1}{4}\omega_{*i}^2 + k_{||}^2 v_A^2 - D} \tag{4.49}$$

$$\gamma = \frac{\omega_r \Gamma}{2\omega_r - \omega_{*i}} \tag{4.50}$$

Here the sign of the denominator in (4.50) is given by the sign chosen for the root in (4.49). If

$$\frac{1}{4}\omega_{*i}^2 + k_{||}^2 v_A^2 > D > k_{||}^2 v_A^2$$

Then the sign of ω_r does not change with the sign of the root. Then we can always find an unstable solution. This is exactly the region in which the MHD instability is stabilised by the FLR effect so the *the dissipative effect restores the stability boundary to that of MHD*. We see, however, that the ion and electron contributions to Γ tend to cancel for $\omega_r \approx \omega_{*i}/2$ so the growthrate may be small in this region.

The kinetic instability in the FLR stabilized region has been verified by linear kinetic calculations Ref. [19]. However, there it was also found that the stability boundary may be lower than the MHD boundary in the presence of magnetic curvature and an ion temperature gradient.

4.6 The Magnetic Drift Mode

For the drift Alfvén wave we noticed that the electromagnetic effects disappeared for $k_{||} = 0$. There is, however, another mode which is electromagnetic and has $k_{||} = 0$. This is the magnetostatic mode which involves only electron motion. The electron motion along B_0 perturbes the magnetic field and the induction force acts back on the electrons. In a homogeneous plasma this mode is purely damped and has zero eigenfrequency. The perturbation of the magnetic field lines form islands in the perpendicular plane and the motion of the electrons along the perturbed field lines causes anomalous heat transfer. In a inhomogeneous plasma, however, this mode has a frequency close to ω_{*e} and is no longer static. the mode is linearly described by a parallel induced electric field and a parallel vector potential $A_{||} = A$ corresponding to a perpendicular magnetic field perturbation. We thus have

$$\mathbf{E} = E_{||}\,\widehat{\mathbf{z}} = -\frac{\partial A}{\partial t}\widehat{\mathbf{z}} \tag{4.51}$$

Again assuming k_x=0 we have

$$\delta\mathbf{B} = \delta B_x\widehat{\mathbf{x}} = ik_yA\widehat{\mathbf{x}} \tag{4.52}$$

Introducing these fields into (4.6) we obtain

$$f_k(\mathbf{v}) = \frac{q}{m}\int_0^\infty i\frac{m}{T}\mathrm{v}_{||}{}'\omega A f_0 \cdot e^{-i\alpha(\tau)}d\tau - \frac{q}{m}\int_0^\infty i\frac{\kappa}{\Omega_c}k_yA(\mathbf{v}\times\widehat{\mathbf{x}})\cdot\widehat{\mathbf{y}}f_0\cdot e^{-i\alpha(\tau)}d\tau \tag{4.53}$$

Which, observing that f_0 is invariant along the orbit, reduces to

$$f_k(\mathbf{v}) = \frac{q}{m}f_0\mathrm{v}_{||}(\omega - \omega_*)A\int_0^\infty e^{-i\alpha(\tau)}d\tau \tag{4.54}$$

Making use of (4.13) we then obtain

$$f_k(\mathbf{v}) = -i\frac{q}{m}f_0\mathrm{v}_{||}(\omega - \omega_*)A\sum_{n,n'}\frac{J_n(\xi)J_{n'}(\xi)e^{-i(n-n')\theta}}{n\Omega_c - \omega} \tag{4.55}$$

Assuming the average $v_{||}$ to be zero, we now find the density perturbation to vanish. The parallel current is

$$j_{||k} = q\int f_k v_{||}\, dv = \frac{q^2 n_0}{T_e}\frac{v_{th}{}^2}{2} A(\omega_{*e} - \omega)\sum_n \frac{\Lambda_n(\varsigma)}{n\Omega - \omega} \tag{4.56}$$

Finally we consider only electron current, take only the term n = 0 and ignore FLR effects. Then using $T_e = m_e v_{the}{}^2/2$ we find

$$j_{||k} = -\frac{e^2 n_0}{m_e}\left(1 - \frac{\omega_{*e}}{\omega}\right)A \tag{4.57}$$

Equation 4.16 for the present case takes the form

$$j_{||k} = -\frac{1}{\mu_0}\frac{\partial^2 A}{\partial y^2} \tag{4.58}$$

Combining (4.57) and (4.58) we have the dispersion relation

$$\omega = \frac{\omega_{*e}}{1 + k_y{}^2 c^2/\omega_{pe}{}^2} \tag{4.59}$$

This is the dispersion relation of the magnetic drift mode in an inhomogeneous plasma. It includes the diamagnetic drift frequency but also the skin depth in the denominator. This is a feature characteristic of including electron inertia.

4.7 The Drift Kinetic Equation

In the limit $k_y{}^2\rho_s{}^2 << 1$ and $\omega << \Omega_c$ the previous procedure of integrating along the Larmor orbits can be avoided. The simplest possible approach in this limit is to write an equation of continuity for guiding centres. Such an equation can be written down immediately once the velocity and acceleration of the guiding centre is known. As it turns out, however, this method requires a more accurate knowledge of the guiding centre dynamics than more systematic procedures starting from the Vlasov equation and it does not give an estimate of the magnitude of the neglected terms. In particular it is difficult to obtain an accurate description of curvature effects. We will thus here restrict ourselves to a slab geometry and leave the more complete description to a later systematic derivation.

The velocity of a guiding centre may be written

$$\mathbf{v}_{gc} = \frac{1}{B}(\mathbf{E} \times \hat{\mathbf{e}}_{||}) + v_{||}\frac{\delta \mathbf{B}_\perp}{\mathbf{B}} + \mathbf{v}_g \tag{4.60}$$

The acceleration is assumed to be directed only parallel to the magnetic field. The continuity equation may as previously be written in the form df/dt = 0 which now becomes

$$\frac{\partial f}{\partial t} + (\mathrm{v}_{\|}\widehat{\mathbf{e}}_{\|} + \mathbf{v_{gc}}) \cdot \nabla f + \frac{q}{m}\left[\mathrm{E}_{\|} + (\mathbf{v_{gc}} \times \delta\mathbf{B}_{\perp}) \cdot \widehat{\mathbf{e}}_{\|}\right]\frac{\partial f}{\partial \mathrm{v}_{\|}} = 0 \tag{4.61}$$

Since (4.61) no longer explicitly depends on $\mathbf{v}_{\perp}$ we may integrate over the perpendicular velocity components. We thus have

$$f = f(\mathbf{r},\mathrm{t},\mathrm{v}_{\|}) \tag{4.62}$$

Equation 4.61 is the simplest form of the drift kinetic equation and does not contain finite Larmor radius effects. It does, however, keep the full parallel kinetic description and can be used to study wave particle resonances. It is a simple exercise to rederive the dispersion relation (4.49) for the magnetic drift mode by using (4.61). A further feature is that (4.61) has not been linearized. Thus it can be used to study nonlinear wave interactions and transport.

4.8 Dielectric Properties of Low Frequency Vortex Modes

We will start by considering flute like modes subject to the condition $\mathrm{k}_{\|}{}^2\mathrm{T_e}/\mathrm{m_i}\omega^2<1$. Dropping the Landau resonance terms but keeping also electron FLR effects we can write the dielectric function observing that $\omega_\mathrm{p}{}^2/\Omega_\mathrm{c}{}^2 = \rho/\Lambda_0{}^2(\mathrm{s_j} = \mathrm{k}^2\mathrm{T_j}/\mathrm{mj}\Omega_\mathrm{cj}{}^2)$

$$\begin{aligned}\varepsilon(k_{\perp},k_{\|},\omega) =&1 + \frac{\omega_{pe}{}^2}{k_{\perp}{}^2\rho_e{}^2\Omega_{ce}{}^2}\left[1 - \frac{\omega-\omega_{*e}}{\omega}\left(1+\frac{k_{\|}{}^2T_e}{m_e\omega^2}\right)\Lambda_0(s_e)\right]\\ &+\frac{\omega_{pi}{}^2}{k_{\perp}{}^2\rho_i{}^2\Omega_{ci}{}^2}\left[1 - \frac{\omega-\omega_{*i}}{\omega}\left(1+\frac{k_{\|}{}^2T_i}{m_i\omega^2}\right)\Lambda_0(s_i)\right]\end{aligned}$$

This expression has several interesting properties which we will investigate. First we expand for small Larmor radius and treat ω_*/ω, $\mathrm{k}_{\|}{}^2\mathrm{T}/\mathrm{m}\omega^2$, and s as small. Then

$$\begin{aligned}\varepsilon(k_{\perp},k_{\|},\omega) =&1 + 2\frac{\omega_{pe}{}^2}{\Omega_{ce}{}^2}\left[\frac{1}{2} + \frac{\kappa}{k_{\perp}}\frac{\Omega_{ce}}{\omega} - \frac{k_{\|}{}^2}{k_{\perp}{}^2}\frac{\Omega_{ce}{}^2}{\omega^2}\right]\\ &+2\frac{\omega_{pi}{}^2}{\Omega_{ci}{}^2}\left[\frac{1}{2} + \frac{\kappa}{k_{\perp}}\frac{\Omega_{ci}}{\omega} - \frac{k_{\|}{}^2}{k_{\perp}{}^2}\frac{\Omega_{ci}{}^2}{\omega^2}\right]\end{aligned} \tag{4.63}$$

Where $\kappa = -\mathrm{dln}\, n_0/dx$.

For a tokamak plasma typically $\omega_{pe} \sim \Omega_{ce}$ while $\omega_{pi} \sim 50\Omega_{ci}$ (we observe also that $\omega_{pi}/\Omega_{ci} = c/v_A$). We notice that the commonly used expression

$$\varepsilon = \varepsilon_F = 1 + \frac{\omega_{pe}{}^2}{\Omega_{ce}{}^2} + \frac{\omega_{pe}{}^2}{\Omega_{ce}{}^2} \tag{4.64}$$

Is usually hard to fulfil in a realistic situation. In cases when the electron contribution can be dropped it may, however, sometimes be fulfilled. Assuming $k_{||} = \kappa = 0$ but keeping the full FLR contribution we obtain

$$\varepsilon(k_\perp) = 1 + \frac{\omega_{pe}{}^2}{\Omega_{ce}{}^2}\frac{1-\Lambda_0(s_e)}{s_e} + \frac{\omega_{pe}{}^2}{\Omega_{ce}{}^2}\frac{1-\Lambda_0(s_i)}{s_i} \tag{4.65}$$

Which shows that ε decreases for large Larmor radius.

The question of quasineutrality is also related to the dielectric constant ε_F. Assuming e.g. that we are in the drift wave region (3.1) and dropping Landau resonances, parallel ion motion and FLR, we obtain from (4.26)

$$\varepsilon(k_\perp, \omega) = 1 + \frac{1}{k_\perp{}^2\lambda_{de}{}^2}\left(1 + k_y{}^2\rho_s{}^2 - \frac{\omega_{*e}}{\omega}\right) \tag{4.66}$$

The dispersion relation for electrostatic drift waves $\varepsilon(k_\perp, \omega) = 0$ can now be written

$$\omega = \frac{\omega_{*e}}{1 + k_y{}^2\rho_s{}^2\left[1 + \lambda_{de}{}^2/\rho_s{}^2\right]} = \frac{\omega_{*e}}{1 + k_y{}^2\rho_s{}^2\left[1 + \Omega_{ci}{}^2/\omega_{pi}{}^2\right]} \tag{4.67}$$

since $\lambda_{de}/\rho_s = \Omega_{ci}/\omega_{pi}$. The condition for applicability of quasineutrality is $k^2\ \lambda_{de}{}^2 << 1$, which leads us to dropping 1 in (4.66). This corresponds to dropping $\Omega_{ci}{}^2/\ \omega_{pi}{}^2$ in (4.67) which is equivalent to assuming that $\varepsilon_F >> 1$. The reason why the condition for quasineutrality can be expressed as $\varepsilon_F >> 1$ without involving the wavelength is that we have compared the deviation from quasineutrality with the ion inertia term $k^2\rho^2$ which also contains the wavenumber.

The wave energy as expressed by the formula

$$W = \frac{1}{4}\omega\frac{\partial}{\partial\omega}\varepsilon(\mathbf{k}, \omega) < E^2 >$$

is closely related to the dielectric properties. We shall here consider the wave energy in two cases.

For electrostatic drift waves, dropping gravity effects but keeping ion FLR effects, we have

$$\varepsilon(k_\perp,\omega) = 1 + \frac{k_{de}^{\,2}}{k_\perp^{\,2}}\left[1 + \frac{T_e}{T_i}\left(1 - \frac{\omega - \omega_{*i}}{\omega}\Lambda_0(s_i)\right)\right] \tag{4.68}$$

Assuming $k_\perp^{\,2}\lambda << 1$(and $\varepsilon_{F.} >> 1$) we can write the dispersion relation

$$\omega = \omega_{*e}\Lambda_0(s_i)\left\{1 + \frac{T_e}{T_i}[1 - \Lambda_0(s_i]\right\}^{-1} \tag{4.69}$$

From (4.68) we also obtain

$$\frac{\partial}{\partial\omega}\varepsilon(\mathbf{k},\omega) = \frac{k_{de}^{\,2}}{k_y^{\,2}}\frac{\omega_{*e}}{\omega^2}\Lambda_0(s_i) \tag{4.70}$$

Inserting (4.69) we then have

$$W_k = \frac{1}{4}k_{de}^{\,2}\left\{1 + \frac{T_e}{T_i}[1 - \Lambda_0(s_i]\right\}|\phi_k|^2 \tag{4.71}$$

Here the second term includes the ion polarization drift and tends to $k^2\ \rho_s^{\,2}$ in the limit $T_e/T_i \to \infty$. For interchange modes ($k_{||} = 0$) we obtain

$$\varepsilon(k_\perp,\omega) = 1 + \frac{k_{de}^{\,2}}{k_\perp^{\,2}}\left[\frac{\omega_{*e}}{\omega} + \frac{T_e}{T_i}\left(1 - \frac{\tilde{\omega} - \omega_{*i}}{\tilde{\omega}}\Lambda_0(s_i)\right)\right] \tag{4.72}$$

Assuming quasineutrality we may write the dispersion relation

$$\frac{\omega_{*e}}{\omega} = -\frac{T_e}{T_i}\left[1 - \frac{\tilde{\omega} - \omega_{*i}}{\tilde{\omega}}\Lambda_0(s_i)\right] \tag{4.73}$$

Multiplying by $\omega - k_y v_g$ we find

$$\frac{k_y v_g\omega_{*e}}{\omega} = \left(\frac{T_e}{T_i}\tilde{\omega} + \omega_{*e}\right)[1 - \Lambda_0(s_i)] \tag{4.74}$$

Alternatively we may write (4.73) as

$$\frac{\omega_{*e}}{\tilde{\omega}}\Lambda_0(s_i) = \frac{\omega_{*e}}{\omega} + \frac{T_e}{T_i}[1 - \Lambda_0(s_i)] \tag{4.75}$$

Differentiating ε we find

$$\frac{\partial\varepsilon}{\partial\omega} = \frac{k_{de}^{\,2}}{k_\perp^{\,2}}\left[\Lambda_0(s_i)\frac{\omega_{*e}}{\tilde{\omega}^2} - \frac{\omega_{*e}}{\omega^2}\right] \tag{4.76}$$

Which, after substitution of (4.75) can be written

$$\frac{\partial \varepsilon}{\partial \omega} = \frac{k_{de}{}^2}{k_\perp{}^2}\frac{1}{\tilde{\omega}}\left[\frac{\omega_{*e} k_y \mathrm{v_g}}{\omega^2} + \frac{T_e}{T_i}[1 - \Lambda_0(s_i)]\right] \tag{4.77}$$

Then using (4.74) we find

$$\frac{\partial \varepsilon}{\partial \omega} = \frac{k_{de}{}^2}{k_\perp{}^2}\frac{1}{\tilde{\omega}}\left[\frac{T_e}{T_i}\frac{\tilde{\omega}}{\omega} + \frac{\omega_{*e}}{\omega} + \frac{T_e}{T_i}\right][1 - \Lambda_0(s_i)] \tag{4.78}$$

From which

$$W = \frac{1}{4}k_{di}{}^2\frac{2\omega - \omega_{*i} - k_y \mathrm{v_g}}{\omega - \mathrm{k_y v_g}}[1 - \Lambda_0(s_i)]|\phi|^2 \tag{4.79}$$

Here we can see that the energy of flute modes is an FLR effect.

4.9 Finite Larmor Radius Effects Obtained by Orbit Averaging

In a fluid description the lowest order finite Larmor radius effects (FLR) can be obtained by including the diamagnetic and stress tensor drifts. Such a calculation, however, becomes rather involved due to cancellation between diamagnetic and stress tensor drifts that are not real particle drifts. Finite Larmor effects are due to the inhomogeneity of the electric field and the correction to the E × B drift caused by it. For a harmonic space dependence of the electric field and $\omega << \Omega_{ci}$ the FLR effect averages the electric field over a range of phases in space and this phase mixing always leads to reduction of the effective field (Fig. 4.2). The efficiency of this phase mixing clearly must depend on the ratio ρ/λ. The particle equation of motion can be written

$$m\frac{d\mathrm{v}}{\mathrm{dt}} = q[\mathbf{E} + \mathbf{v} \times \mathbf{B}] \tag{4.80}$$

For simplicity we use a slab geometry according to Fig. 4.3, where $\mathbf{B} = \mathrm{B}_0\hat{\mathbf{z}}$ and

$$\mathbf{E} = \mathrm{E}_0 \cos(ky - \omega t)\hat{x} \tag{4.81}$$

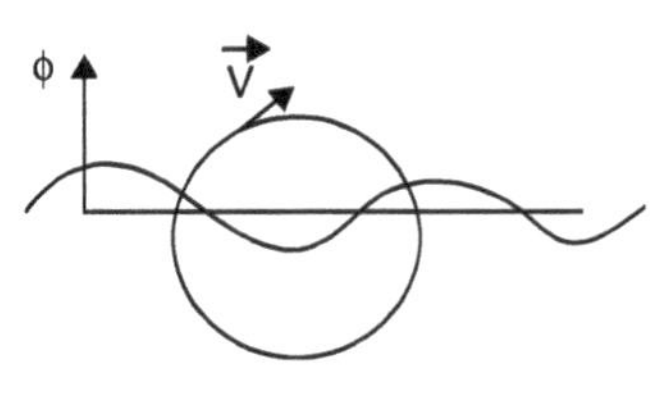

Fig. 4.2 Finite gyroradius averaging

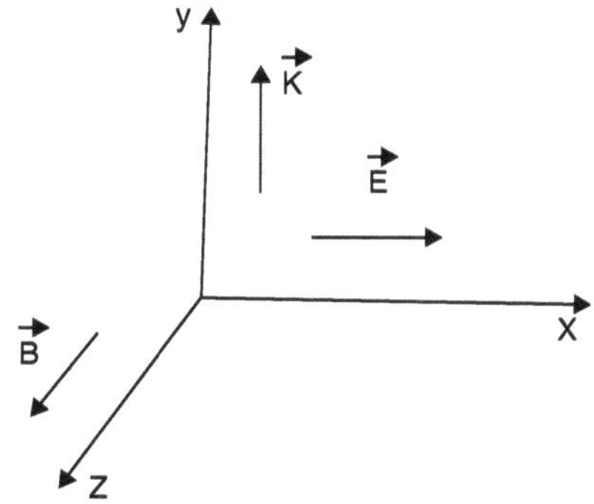

Fig. 4.3 Slab geometry with electric field

In component form we have

$$\frac{dv_x}{dt} = \Omega_c v_y + \frac{q}{m} E_0 \cos(ky(t) - \omega t) \tag{4.82}$$

$$\frac{dv_y}{dt} = -\Omega_c v_x \tag{4.83}$$

where we observe that the electric field is evaluated at y(t), i.e. along the orbit. The coupling between the equations on the time scale Ω_c^{-1} can be eliminated by differentiating with respect to t and substitution. This leads to

$$\frac{d^2 v_x}{dt^2} = -\Omega_c^2 v_x + \frac{q}{m}\left(\omega - k\frac{dy}{dt}\right) E_0 \sin[ky(t) - \omega t] \tag{4.84}$$

$$\frac{d^2 v_y}{dt^2} = -\Omega_c^2 v_y - \Omega_c^2 \frac{E_0}{B_0} \cos[ky(t) - \omega t] \tag{4.85}$$

We shall now assume that $\Omega_c >> \omega$ so that the time scales are well separated. We then average over the short timescale obtaining

$$<v_x> = \frac{E_0}{B_0 \Omega_c}\left\langle \left(\omega - k\frac{dy}{dt}\right) \sin[ky(t) - \omega t] \right\rangle - \frac{1}{\Omega_c^2}\left\langle \frac{d^2 v_x}{dt^2} \right\rangle \tag{4.86}$$

$$<v_y> = -\frac{E_0}{B_0}\langle \cos[ky(t) - \omega t] \rangle - \frac{1}{\Omega_c^2}\left\langle \frac{d^2 v_y}{dt^2} \right\rangle \tag{4.87}$$

We shall now perform the averaging of (4.86) and (4.87) over the unperturbed orbit, obtained by solving (4.68) with $E_0 = 0$. This orbit may be written

$$y(t) = y_0 + r_L(t) \tag{4.88}$$

Where

$$r_L(t) = -\frac{v_\perp}{\Omega}[\cos(\Omega_c t + \phi) - \cos\phi] \tag{4.89}$$

Is the projection of the Larmor radius along y. The orbit in (4.89) corresponds to

$$v_y = v_\perp \sin(\Omega_c t + \phi)$$

$$v_x = v_\perp \cos(\Omega_c t + \phi)$$

For the orbit (4.89) we have $<d^2v_{x,y}/dt^2> = 0$. we are also interested in the lowest order FLR effects and take only linear terms in the parameter $k^2 r_L{}^2 << 1$. We then have

$$\begin{aligned}\sin(ky(t) - \omega t) =& \sin(ky_0 - \omega t)\left\{1 - \frac{1}{2}\frac{k^2 v_\perp{}^2}{\Omega_c{}^2}[\cos(\Omega_c t + \phi) - \cos\phi]^2\right\}\\ &- \cos(ky_0 - \omega t)\frac{kv_\perp}{\Omega_c}[\cos(\Omega_c t + \phi) - \cos\phi]\\ \cos(ky(t) - \omega t) =& \cos(ky_0 - \omega t)\left\{1 - \frac{1}{2}\frac{k^2 v_\perp{}^2}{\Omega_c{}^2}[\cos(\Omega_c t + \phi) - \cos\phi]^2\right\}\\ &+ \sin(ky_0 - \omega t)\frac{kv_\perp}{\Omega_c}[\cos(\Omega_c t + \phi) - \cos\phi]\end{aligned}$$

We now perform the averaging over time, keeping ωt constant since $\Omega_c >> \omega$. We then obtain

$$\begin{aligned}< \sin(ky(t) - \omega t)> =& \sin(ky_0 - \omega t)\left\{1 - \frac{1}{2}\frac{k^2 v_\perp{}^2}{\Omega_c{}^2}\left[\frac{1}{2} + \cos^2\phi\right]\right\}\\ &+ \cos(ky_0 - \omega t)\frac{kv_\perp}{\Omega_c}\cos\phi\end{aligned} \tag{4.90}$$

$$\begin{aligned}< \cos(ky(t) - \omega t)> =& \cos(ky_0 - \omega t)\left\{1 - \frac{1}{2}\frac{k^2 v_\perp{}^2}{\Omega_c{}^2}\left[\frac{1}{2} + \cos^2\phi\right]\right\}\\ &- \sin(ky_0 - \omega t)\frac{kv_\perp}{\Omega_c}\cos\phi\end{aligned} \tag{4.91}$$

We also need

$$\left\langle \frac{d}{dy}\sin(ky_0 - \omega t)\right\rangle = 0$$

We now continue to average (4.91) over a Maxwellian distribution. Then $v_\perp{}^2 = 2(T/m)$ $\cos\varphi = 0$, $\cos^2\varphi = 1/2$ and

$$<<\sin(ky(t) - \omega t)>> = \sin(ky_0 - \omega t)\left(1 - k^2 \frac{T}{m\Omega_c^2}\right)$$

$$<<\cos(ky(t) - \omega t)>> = \cos(ky_0 - \omega t)\left(1 - k^2 \frac{T}{m\Omega_c^2}\right)$$

Introducing $\rho^2 = 2T/(m\Omega_c{}^2)$ we now have

$$<<v_x>> = \frac{\omega E_0}{B_0\Omega_c}\sin(ky_0 - \omega t)\left(1 - \frac{1}{2}k^2\rho^2\right) \tag{4.92}$$

$$<<v_y>> = \frac{E_0}{B_0}\cos(ky_0 - \omega t)\left(1 - \frac{1}{2}k^2\rho^2\right) \tag{4.93}$$

Here the gyroradius ρ is for a general species. However, the application will almost exclusively be to the main ions. These are the averaged drifts we have been seeking. With the present choice of **E** and **k** we have the **E** × **B** drift in the y direction and the polarisation drift in the x direction. As it turns out, the present averaging is not accurate enough to give correct FLR correction to the polarisation drift. Thus if the perturbed orbit is introduced into $<d^2v_{x,y}/dt^2>$, we obtain new terms of the same order as the FLR correction to (4.92). Neglecting FLR corrections to the polarization drift we obtain in vector form

$$<\mathbf{v}_\perp> = \left(1 - \frac{1}{2}k^2\rho^2\right)\mathbf{v}_\mathbf{E} + \mathbf{v}_\mathbf{p} \tag{4.94}$$

Where we used only one averaging sign referring to both time and velocity averagings.

We observe here that the constant of integration, cosφ, in (4.89) is important in order to include all particles with orbits through y_0. Since $y = y_0 + (v_\perp/\Omega_c)\cos\varphi$ the representation (4.89) means that we include particles with gyrocentres between $y_0 - v_\perp/\Omega_c$ and $y_0 + v_\perp/\Omega_c$.

In order to compare our results with those from a fluid theory we now calculate a density response to an electric field by using the continuity equation. This is a natural procedure since the density response is uniquely defined while fluid and particle drifts may differ. Thus using (4.94) for ions in the continuity equation, neglecting parallel ion motion, we obtain for $k_\perp >> |d \ln n_0/dx|$

$$\frac{\delta n}{n_0} = \left[\frac{\omega_{*e}}{\omega}\left(1 - \frac{k_\perp{}^2 T_i}{m_i\Omega_{ci}{}^2}\right) - \frac{k_\perp{}^2 T_e}{m_i\Omega_{ci}{}^2}\right]\frac{e\phi}{T_e} \tag{4.95}$$

Where we included the ion FLR correction to the E × B drift according to (4.94) and the polarization drift. Equation 4.95 can also be rewritten in the form

$$\frac{\delta n}{n_0} = \left[\frac{\omega_{*e}}{\omega} - \frac{k_\perp{}^2 T_e}{m_i \Omega_{ci}{}^2}\left(1 - \frac{\omega_{*i}}{\omega}\right)\right]\frac{e\phi}{T_e} \tag{4.96}$$

Where the FLR term now appears as the $-\omega_{*i}/\omega$ correction to the polarisation drift. Equation 4.96 can be obtained by using fluid equations and including the diamagnetic drift in the convective derivative in the polarisation drift, i.e.

$$\frac{\partial}{\partial t} \to \frac{\partial}{\partial t} + \mathbf{v_{*j}} \cdot \nabla$$

This is also what remains in the fluid description after cancellations between diamagnetic and stress tensor drifts (compare Chap. 2). It can also be readily shown by the orbit averaging method that this procedure also can be used for the perturbed diamagnetic drift, thus giving the lowest order nonlinear FLR effects when used in the convective derivative in the polarisation drift.

It is also interesting to note the similarity between the FLR effects and the polarisation drift in their contribution to the density response. Such a similarity may be expected since the FLR effect is due to the space dependence of the electric field along the orbit while the polarisation drift is due to the time dependence. A particle gyrating in the orbit cannot distinguish between these origins of field variation.

4.10 Discussion

We have in this chapter rederived the most important dispersion relations of Chap. 3 using a kinetic description. This has been simplified by using a slab geometry. A more general Gyro-kinetic description will be given in Chap. 5. We have also particularly considered the effects of finite Larmor radius and verified the first order effects that were obtained from fluid theory in Chap. 2 and the consequences of it for stability found in Chap. 3. Finally the dielectric properties of inhomogeneous plasmas are fundamental. We will later, in Chap. 7, show how the wave energy of interchange modes can be recovered to first order in the FLR parameter from a nonlinear conservation relation. In Chap. 6 we will use more realistic geometries and also study modes driven by temperature gradients.

4.11 Exercises

1. Perform the integration in (4.3).
2. Derive a relation between φ and $A_{||}$ in (4.7) using fluid equations and neglecting parallel ion motion. Compare with the expression for $E_{||}$ in (3.71)

3. Derive (4.47) by using fluid equations.
4. Derive a dispersion relation for the 'Universal drift instability for small Larmor radius by using (4.22) for electrons and fluid equations for ions, i.e. neglecting ion Landau damping.
5. Use the tokamak data in Appendix I to compare the different contributions to ε_F in (4.64).

References

1. M.N. Rosenbluth and C. Longmire, Ann. Phys. (N.Y.) **1**, 120 (1957).
2. M.N. Rosenbluth and N. Rostoker, Phys. Fluids **2**, 23 (1959).
3. L.I. Rudakov and R.Z. Sagdeev Sov. Phys. JETP **37,** 952 (1960).
4. M.N. Rosenbluth, N.A. Krall and N.Rostoker, Nucl. Fusion Suppl. Pt. **1** 143 (1962).
5. N. A. Krall and M.N. Rosenbluth, Phys. Fluids **5**, 1435 (1962).
6. B. B. Kadomtsev, Plasma Turbulence, Academic Press, New York 1965
7. A.G. Sitenko, Electromagnetic Fluctuations in a Plasma, Academic Press, N.Y. 1967.
8. N. A. Krall in Advances in Plasma Physics (Ed. A. Simon and W. Thompson), Wiley, New York Vol. **1**, p. 153 (1968)
9. B. Coppi, G. Laval, R. Pellat and M.N. Rosenbluth, Plasma Physics **10** 1 (1968).
10. L.I. Rukhadze and V.P. Silin, Soviet Physics Usp. **2**, 659 (1969).
11. B. B. Kadomtsev and O.P. Pogutse, in Reviews of Plasma Physics (Ed. M.A. Leontovitch) Consultant Bureau, New York, Vol. **5**, p. 249 (1970).
12. S. Ichimaru, Basic Principles of Plasma Physics, Mc Graw Hill 1973.
13. N. A. Krall and Trivelpiece, Principle of Plasma Physics, McGraw Hill 1973.
14. A.B. Mikhailovskii, Theory of Plasma Instabilities, Vol. **2**, Consultant Bureau, New York 1974.
15. A. Hasagawa, Plasma Instabilities and Nonlinear effects, Springer 1975, Chapter 3.
16. W.M. Manheimer, An Introduction to trapped Particle Instabilities in Tokamak, ERDA Crit. Rev. Series 1977.
17. S.P. Gary and J.J. Sanderson, Phys. Fluids **21**, 1181 (1978).
18. A. Hasegawa, Phys. Fluids **22**, 1988 (1979).
19. C.Z. Cheng, Phys. Fluids **25**, 1020 (1982)

Chapter 5
Kinetic Descriptions of Low Frequency Modes Obtained by Gyroaveraging

We have, in Chap. 4 studied kinetic descriptions in simple geometries. Characteristic of these has been that inhomogeneities have been assumed to be constant along particle orbits. This can be achieved by representing magnetic curvature by a simple gravity force. Then, it is possible to integrate the Vlasov equation in a magnetized plasma, along the characteristics (linear orbit) for all times. This can be made for arbitrary frequencies and gyroradii, thus including cyclotron resonances and the full Finite Larmor Radius (FLR) effects. This can also be done keeping nonlinear terms although we only did that for the drift kinetic equation which does not involve FLR effects. In the present chapter we will drop the assumption of inhomogeneities that are constant in space and include the full kinetic magnetic drifts [1–20]. In this case we do not know even the unperturbed orbits for all times. This case can still be treated in a reasonably simple form if we restrict our study to low frequency modes which have $\omega << \Omega_c$. We can then average over the fast timescale. We will start with the simplest case when FLR effects are small and derive a more general drift kinetic equation than we did in Chap. 4. We also include a brief survey of this area [1–20].

5.1 The Drift Kinetic Equation

The complexity of a full Vlasov description in a magnetised plasma has led to the development of a number of simplified approximate descriptions in various limits. One obvious limit is the case of strongly magnetized particles [16, 20]. In this limit the particles are well localized in the plane perpendicular to the magnetic field so that the kinetic description is needed only along the field. This approximation is usually valid for electrons in laboratory plasmas and sometimes also for ions. The condition for localization in the perpendicular plane may be written $\rho \ll \lambda$ where ρ is the Larmor radius and λ is the inhomogeneity scale length of the phenomenon we want to study.

J. Weiland, *Stability and Transport in Magnetic Confinement Systems*, Springer Series on Atomic, Optical, and Plasma Physics 71,
DOI 10.1007/978-1-4614-3743-7_5, © Springer Science+Business Media New York 2012

Another simplification that still leaves a large class of important phenomena within the description is to assume that the time scale of the phenomenon we are interested in is much longer than the gyroperiod, i.e $\omega \ll \Omega_{ci}$. When these conditions are fulfilled it is easy to average the Vlasov equation over a Larmor orbit since both the distribution function and the electromagnetic fields are almost constants over the Larmor orbit in this case. We will, however, include magnetic curvature, thus keeping background drifts proportional to ρ/L_B where L_B is the inhomogeneity scale length of the background magnetic field. This effect may sometimes be more important than finite Larmor radius (FLR) effects of order ρ/λ since it enters multiplied by the large scale thermal velocity.

We shall here use a *Lagrangian method of averaging*, i.e. we follow the particle around the gyro orbit instead of averaging over the short time-scale at a point. The averaging procedure is then considerably simplified since we already know the average of the particle velocity, i.e. the guiding center drift. We thus have:

$$<\mathbf{v}_\perp> = \mathbf{v_E} + v_{||}\frac{\delta B_\perp}{B} + \mathbf{v_D} \tag{5.1}$$

Where

$$\mathbf{v_E} = \frac{1}{B_0}(\mathbf{E} \times \hat{\mathbf{e}}_{||})$$

$$\mathbf{v}_D = \mathbf{v}_\kappa + \mathbf{v}_{\nabla \mathbf{B}}$$

$$\mathbf{v}_\kappa = \frac{v_{||}{}^2}{\Omega_c}(\hat{\mathbf{e}}_{||} \times \boldsymbol{\kappa})$$

$$\mathbf{v}_{\nabla B} = \frac{v_\perp{}^2}{2\Omega_c}(\hat{\mathbf{e}}_{||} \times \nabla \ln B)$$

$$\boldsymbol{\kappa} = (\hat{\mathbf{e}}_{||} \cdot \nabla)\hat{\mathbf{e}}_{||}$$

E is the electric field, $\mathbf{B_0}$ is the background magnetic field and $\delta\mathbf{B}_\perp$ is the perpendicular magnetic field perturbation.

We start from the Vlasov equation in the form

$$\frac{\partial f}{\partial t} + \mathbf{v} \cdot \nabla f + \frac{q}{m}(\mathbf{E} + \mathbf{v} \times \mathbf{B}) \cdot \frac{\partial f}{\partial \mathbf{v}} = 0 \tag{5.2}$$

We now separate our description into the directions parallel and perpendicular to $\mathbf{B_0}$, using the notations $\|$ and $\perp$ respectively. The velocity is then written

$$\mathbf{v} = \mathbf{v}_\perp + v_{||}\hat{\mathbf{e}}_{||}$$

Where $\mathbf{e}_{||} = \mathbf{B}_0/\mathrm{B}$ and the angle is given by $\varphi = \widehat{\mathbf{e}}_{||} \times \mathbf{v}_{\perp}$.

The velocity gradient may then be written

$$\frac{\partial}{\partial \mathbf{v}} = \widehat{\mathbf{v}}_{\perp} \frac{\partial}{\partial \mathrm{v}_{\perp}} + \widehat{\phi} \frac{1}{\mathrm{v}_{\perp}} \frac{\partial}{\partial \phi} + \widehat{\mathbf{e}}_{||} \frac{\partial}{\partial \mathrm{v}_{||}}$$

and then

$$(\mathbf{v} \times \mathbf{B}) \cdot \frac{\partial}{\partial \mathbf{v}} = (\widehat{\mathbf{v}}_{\perp} \times \mathbf{B}_{||}) \cdot \widehat{\phi} \frac{1}{\mathrm{v}_{\perp}} \frac{\partial}{\partial \phi} + (\mathbf{v} \times \delta \mathbf{B}_{\perp}) \cdot \frac{\partial}{\partial \mathbf{v}}$$

Equation 5.2 then reduces to

$$\frac{\partial f}{\partial t} + \mathbf{v} \cdot \nabla f + \frac{q}{m} (\mathbf{E} + \mathbf{v} \times \delta \mathbf{B}_{\perp}) \cdot \frac{\partial f}{\partial v} - \frac{B_{||}}{\mathrm{B}} \Omega_c \frac{\partial f}{\partial \phi} = 0 \tag{5.3}$$

where the factor $\mathbf{B}_{||}/\mathrm{B}$ accounts for perturbations in B parallel to the background field. This factor is always of order 1.

We shall now use the assumption that Ω_c is much larger than any other frequency in (5.3). To lowest order in Ω_c^{-1} (5.3) then leads to the condition

$$\frac{\partial f}{\partial \phi} = 0$$

This means that we as a first approximation can treat f as independent of φ in all terms except the last term in (5.3). We now, however, also want to keep curvature terms proportional to ρ/Ω_c. These terms are first order in Ω_c^{-1} so some care is needed in treating them. We shall assume that $f = f(\mathrm{t}, r, \mathrm{v}_{||}^2, \mathrm{v}_{\perp}^2, \varphi)$. The most important curvature dependence of f enters in the separation between $\mathrm{v}_{||}$ and $\mathbf{v}_{\perp}$. If we separate out this additional space dependence we may write

$$\nabla f = \frac{\partial f}{\partial \mathbf{r}} + \frac{\partial f}{\partial \mathrm{v}_{\perp}^2} \nabla \mathrm{v}_{\perp}^2 + \frac{\partial f}{\partial \mathrm{v}_{||}^2} \nabla \mathrm{v}_{||}^2$$

Or

$$\nabla f = \frac{\partial f}{\partial \mathbf{r}} + \mathrm{v}_{||} \nabla (\widehat{\mathbf{e}}_{||} \cdot \mathbf{v}) \left(\frac{\mathbf{1}}{\mathrm{v}_{||}} \frac{\partial f}{\partial \mathrm{v}_{||}} - \frac{1}{\mathrm{v}_{\perp}} \frac{\partial f}{\partial \mathrm{v}_{\perp}} \right) \tag{5.4}$$

where the space dependence of $\mathbf{e}_{||}$ has been separated out. It can now be shown that

$$\mathbf{v} \cdot \nabla (\widehat{\mathbf{e}}_{||} \cdot \mathbf{v}) = \mathrm{v}_{||} (\mathbf{v} \cdot \boldsymbol{\kappa}) \tag{5.5}$$

and we notice that this curvature depends on the phase angle φ.

We now perform the gyroaveraging of (5.2). Then $\partial f/\partial t$ is unchanged. In the second term $v_{||}e_{||} \bullet(\partial f/\partial \mathbf{r})$ is unchanged while $\mathbf{v}$ is replaced by $\mathbf{v}_{gc}$ in the perpendicular part. In the third term we can neglect $\partial f/\partial \varphi$. The part containing $\partial f/\partial v_\perp$ reduces to

$$\frac{q}{m}\left[\mathbf{E} + v_{||}(\widehat{\mathbf{e}}_{||} \times \delta\mathbf{B}_\perp)\right] \cdot \frac{\mathbf{v}_\perp}{v_\perp}\frac{\partial f}{\partial \mathbf{v}_\perp}$$

Here the averaging leads to a replacement of $\mathbf{v}_\perp$ by $\mathbf{v}_{gc}$. As it turns out, however, the first two parts of $\mathbf{v}_{gc}$ do not contribute due to orthogonality so that we are left with only $\mathbf{v}_D$. In the part containing $\partial f/\partial v_{||}$ we may just replace $\mathbf{v}_\perp$ by $\mathbf{v}_{gc}$. The last term in (5.2) finally is a total derivative in φ and vanishes since orbit averaging means integrating one period in φ.

Thus writing down our averaged equation directly as we obtain it after orbit averaging we have

$$\begin{aligned}
&\frac{\partial f}{\partial t} + (v_{||}\widehat{\mathbf{e}}_{||} + \mathbf{v}_{gc}) \cdot \frac{\partial f}{\partial \mathbf{r}} + v_{||}{}^2 \boldsymbol{\kappa} \cdot \mathbf{v}_{gc}\left(\frac{1}{v_{||}}\frac{\partial f}{\partial v_{||}} - \frac{1}{v_\perp}\frac{\partial f}{\partial v_\perp}\right)\\
&+\frac{q}{m}\left[E_{||} + (\mathbf{v}_{gc} \times \delta\mathbf{B}_\perp) \cdot \widehat{\mathbf{e}}_{||}\right]\frac{\partial f}{\partial v_{||}}\\
&+\frac{q}{m}\left[\mathbf{E} + v_{||}(\widehat{\mathbf{e}}_{||} \times \delta\mathbf{B}_\perp)\cdot\right]\frac{\mathbf{v}_D}{v_\perp}\frac{\partial f}{\partial v_\perp} = 0
\end{aligned} \tag{5.6}$$

Since now

$$\boldsymbol{\kappa} \cdot \mathbf{v}_{gc} = \frac{1}{B}(\widehat{\mathbf{e}}_{||} \times \boldsymbol{\kappa}) \cdot \mathbf{E} + \frac{v_{||}}{B}(\widehat{\mathbf{e}}_{||} \times \boldsymbol{\kappa}) \cdot (\widehat{\mathbf{e}}_{||} \times \delta\mathbf{B})$$

the third term may be written as

$$\frac{q}{m}\mathbf{v}_\kappa \cdot \left[\mathbf{E} + v_{||}(\widehat{\mathbf{e}}_{||} \times \delta\mathbf{B})\right]\left(\frac{1}{v_{||}}\frac{\partial f}{\partial v_{||}} - \frac{1}{v_\perp}\frac{\partial f}{\partial v_\perp}\right)$$

We then obtain the drift kinetic equation

$$\begin{aligned}
&\frac{\partial f}{\partial t} + (v_{||}\widehat{\mathbf{e}}_{||} + \mathbf{v}_{gc}) \cdot \frac{\partial f}{\partial \mathbf{r}} + \frac{q}{m}\left[E_{||} + (\mathbf{v}_{gc} \times \delta\mathbf{B}_\perp) \cdot \widehat{\mathbf{e}}_{||}\right]\frac{\partial f}{\partial v_{||}}\\
&+\frac{q}{m}\left[\mathbf{E} + v_{||}(\widehat{\mathbf{e}}_{||} \times \delta\mathbf{B})\right] \cdot \left(\frac{\mathbf{v}_{\nabla B}}{v_\perp}\frac{\partial f}{\partial v_\perp} + \frac{\mathbf{v}_\kappa}{v_{||}}\frac{\partial f}{\partial v_{||}}\right) = 0
\end{aligned} \tag{5.7}$$

We notice that the first three terms can easily be obtained from a continuity equation for guiding centers (cf Eq. 4.51). Equation 5.7 agrees to first order in the inverse aspect ratio with the drift kinetic equation derived by D'Ippolito and Davidson [6.20] except for the presence of the $\mathbf{v}_D \bullet \partial f/\partial \mathbf{r}$ term in (5.7). This term is comparable to the other curvature terms if $f_1 \sim q\varphi/T$ and is usually kept.

If we replace the dependence on $v_{\|}{}^2$ and $v_{\perp}{}^2$ of f by $E = v^2/2$ and $\mu - (\partial\mu/\partial r)r$ where $\mu = v_{\perp}{}^2/2B$, and r is the perpendicular direction of gradB_0, we can obtain also the mirror force terms kept by D'Ippolito and Davidson. The correction $-(\partial\mu/\partial r)r$ of μ is necessary in order to have conservation to lowest order on the $\Omega_c{}^{-1}$ time scale. We then obtain a correction to (5.4) of the form $(\partial f/\partial\mu)\mathrm{grad}_{\|}\mu$

i.e

$$\frac{1}{2}v_{\perp}{}^2 \cdot \left(\frac{\mathbf{1}}{v_{\|}}\frac{\partial f}{\partial v_{\|}} - \frac{1}{v_{\perp}}\frac{\partial f}{\partial v_{\perp}}\right)\widehat{\mathbf{e}}_{\|} \cdot \nabla \ln B \tag{5.8}$$

The drift kinetic equation (5.7) has here been obtained in a comparatively simple way. It does not take into account finite Larmor radius effects of the type $k^2\rho^2$ but includes the full parallel dynamics, is fully nonlinear and makes no WKB assumption for the space scale of perturbations.

5.1.1 *Moment Equations*

In order to see what fluid motion it corresponds to we shall now take moments of (5.7). This procedure is rather complicated in the presence of $\mathbf{v}_D$ which in general depends on both $v_\perp$ and $v_\|$. For this reason we will in the following for simplicity neglect curvature effects.

The zeroth moment is then

$$\frac{\partial n}{\partial t} + \widehat{\mathbf{e}}_{\|} \cdot \nabla(n u_{\|}) + \int \mathbf{v_{gc}} \cdot \nabla f d\mathbf{v} = 0$$

where **u** is the fluid velocity. Now inserting (5.1) where $\mathbf{v}_D = 0$,, we obtain the continuity equation

$$\frac{\partial n}{\partial t} + \frac{1}{q}\left[\frac{\partial}{\partial z}j_{\|} + \frac{\delta\mathbf{B}_{\perp}}{\mathbf{B_0}} \cdot \nabla j_{\|}\right] + \frac{1}{B_0}(\mathbf{E} \times \widehat{\mathbf{e}}_{\|}) \cdot \nabla n = 0 \tag{5.9}$$

where $j_\|$ is the parallel current. The first parallel moment of (5.7) may be written:

$$\frac{\partial}{\partial t}(n u_{\|}) + \widehat{\mathbf{e}}_{\|} \cdot \nabla \int v_{\|}{}^2 f dv_{\|} + \int \mathbf{v}_{gc} \cdot \nabla f v_{\|} dv_{\|} - \frac{q}{m}\left[E_{\|} + (\mathbf{v}_E \times \delta\mathbf{B}_{\perp}).\widehat{\mathbf{e}}_{\|}\right] n = 0$$

where

$$\mathbf{v_E} = \frac{1}{B_0}(\mathbf{E} \times \widehat{\mathbf{e}}_{\|})$$

Now $v_{\|} = u_{\|} + w_{\|}$ where $w_\|$ is the thermal random velocity. Thus

$$\int v_{\|}{}^2 f dv_{\|} = \int u_{\|}{}^2 f dv_{\|} + \int w_{\|}{}^2 f dw_{\|} = n u_{\|}{}^2 + \frac{1}{n}P$$

where P is the pressure. Substituting now (5.9) for $\partial n/\partial t$ we obtain

$$n\frac{\partial}{\partial t}u_{||} + nu_{||}\widehat{\mathbf{e}}_{||}\cdot\nabla u_{||} + n\frac{1}{\mathrm{B}_0}(\mathbf{E}\times\widehat{\mathbf{e}}_{||})\cdot\nabla u_{||} + nu_{||}\frac{\delta\mathbf{B}_\perp}{B}\cdot\nabla u_{||}$$
$$+\frac{1}{m}\widehat{\mathbf{e}}_{||}\cdot\nabla P + \frac{1}{m}\frac{\delta\mathbf{B}_\perp}{B}\cdot\nabla P - \frac{q}{m}\left[\mathrm{E}_{||} + (\mathbf{v}_{\mathrm{E}}\times\delta\mathbf{B}_\perp).\widehat{\mathbf{e}}_{||}\right]n = 0 \quad (5.10)$$

Equation 5.10 is the parallel equation of motion in the absence of FLR effects. It is important to note here the absence of diamagnetic drifts in the convective part of the time derivative. It is instructive to rewrite (5.7) slightly. We may define the perpendicular guiding center fluid velocity

$$\mathbf{u}_{\mathrm{gc}} = \frac{1}{\mathrm{B}_0}(\mathbf{E}\times\widehat{\mathbf{e}}_{||}) + u_{||}\frac{\delta\mathbf{B}_\perp}{\mathbf{B_0}}$$

Introducing now the diamagnetic drift velocity

$$\mathbf{v}_* = \frac{1}{\mathrm{qnB}_0}(\widehat{\mathbf{e}}_{||}\times\nabla P)$$

we have

$$\frac{q}{m}(\mathbf{v}_*\times\delta\mathbf{B})\cdot\widehat{\mathbf{e}}_{||} = \frac{q}{m}(\widehat{\mathbf{e}}_{||}\times\mathbf{v}_*)\cdot\delta\mathbf{B}_\perp = -\frac{1}{mn}\frac{\delta\mathbf{B}_\perp}{B}\cdot\nabla P$$

We may thus rewrite (5.10) in form

$$\frac{\partial}{\partial t}u_{||} + u_{||}\widehat{\mathbf{e}}_{||}\cdot\nabla u_{||} + \mathbf{u}_{\mathrm{gc}}\cdot\nabla u_{||} = \frac{q}{m}\left\{\mathrm{E}_{||} + [(\mathbf{v}_{\mathrm{E}} + \mathbf{v}_*)\times\delta\mathbf{B}_\perp]\cdot.\widehat{\mathbf{e}}_{||}\right\} - \frac{1}{mn}\widehat{\mathbf{e}}_{||}\cdot\nabla P \quad (5.11)$$

Equation 5.11is the usual parallel equation of motion where the diamagnetic drift is included in the $\mathbf{v}\times\mathbf{B}$ term but not in the convective derivatives.

5.1.2 The Magnetic Drift Mode

We now restrict our consideration to the case $\mathbf{e}_{||}\bullet\mathrm{grad} = 0$ and linearize. Equation 5.11 is then

$$\frac{\partial}{\partial t}u_{||} = -\frac{e}{m}\mathrm{E}_{||} - \frac{e}{m}(\mathbf{v}_*\times\delta\mathbf{B}_\perp)\cdot.\widehat{\mathbf{e}}_{||}$$

Using the representation $\mathbf{A} = A\hat{\mathbf{e}}_{||}$ and

$$\delta\mathbf{B}_\perp = \nabla \times A\hat{\mathbf{e}}_{||} = \nabla_\perp A \times \hat{\mathbf{e}}_{||} \quad \mathbf{E} = -\nabla\phi - \frac{\partial A}{\partial t}\hat{\mathbf{e}}_{||}$$

We obtain

$$-i\omega u_{||} = -i\omega\frac{e}{m}A - i\frac{e}{m}\mathbf{v}_* \cdot \left[(\mathbf{k} \times \hat{\mathbf{e}}_{||}) \times \hat{\mathbf{e}}_{||}\right]A$$

Or

$$u_{||} = \frac{e}{m}A\left(1 - \frac{\omega_{*e}}{\omega}\right)$$

The parallel current is

$$j_{||} = \frac{1}{\mu}(\nabla \times \mathbf{B}) \cdot \hat{\mathbf{e}}_{||} = \frac{1}{\mu}\left[\nabla \times (\nabla_\perp A \times \hat{\mathbf{e}}_{||})\right] \cdot \hat{\mathbf{e}}_{||} = -\frac{1}{\mu}\Delta_\perp A$$

Thus

$$neu_{||} = -\frac{ne^2}{m}A\left(1 - \frac{\omega_{*e}}{\omega}\right) = \frac{1}{\mu}k_\perp{}^2 A$$

or

$$\omega = \omega_{*e}\left(1 - \frac{k_\perp{}^2 c^2}{\omega_{pe}{}^2}\right)^{-1}$$

This is the dispersion relation of the magnetic drift mode (compare Eq. 4.59). We see here that the inclusion of the diamagnetic drift in the convective derivative would not cause a negligible modification. We must then conclude that this term is cancelled by stress tensor effects.

Transport due to an enhanced thermal equilibrium spectrum of magnetic drift modes is discussed in Sect. 9.2.

5.1.3 The Tearing Mode

In a more realistic geometry with magnetic shear, the condition $k_{||} = 0$, which was assumed for the magnetic drift mode, can not be fulfilled everywhere for modes with finite radial extent. We thus have to solve a radial eigenvalue problem. The characteristic property $\varphi = 0$ of the magnetic drift mode will then enter as a boundary condition at the rational surface. This type of mode can be driven unstable by collisions and is called the tearing mode [6.4]. In a shortwave version it is called the microtearing mode [6.57, 6.58].

5.2 The Linear Gyrokinetic Equation

We have in Chap. 4 seen how we can obtain general kinetic equations in a slab geometry by integrating along unperturbed orbits. For more general geometries we have in the previous section derived a drift kinetic equation valid in the limit $k_\perp \rho \ll 1$. We will here finally consider the case $k_\perp \rho \sim 1$ in complex geometry. Gyroaveraged equations of this type are called Gyrokinetic equations. A pioneering work along these lines is that by Rutherford and Frieman [1]. A generally used assumption in this type of equations is: $\rho/L \ll 1$ where L is an equilibrium scale-length. We will here derive gyrokinetic equations by a method that is considerably shorter than conventional methods. We may write the Vlasov equation in the form:

$$\frac{Df}{Dt} = 0 \tag{5.12}$$

where

$$\frac{D}{Dt} = \frac{\partial}{\partial t} + \mathbf{v} \cdot \frac{\partial}{\partial \mathbf{r}} + \frac{q}{m}(\mathbf{E} + \mathbf{v} \times \mathbf{B}) \cdot \frac{\partial}{\partial \mathbf{v}} \tag{5.13}$$

We shall assume a solution of the form:

$$f(\mathbf{r}, \mathbf{v}, t) = f_0(\mathbf{r}, \mathbf{v}) + f_1(\mathbf{r}, \mathbf{v}, t)$$

Where f_0 is the background distribution and f_1 is a perturbation fulfilling

$$f_1(\mathbf{r}, \mathbf{v}, t) << f_0(\mathbf{r}, \mathbf{v})$$

For simplicity we shall here omit background electric fields. The equation for f_0 becomes

$$\mathbf{v} \cdot \frac{\partial f_0}{\partial \mathbf{r}} + \frac{q}{m}(\mathbf{v} \times \mathbf{B}_0) \cdot \frac{\partial f_0}{\partial \mathbf{v}} = 0 \tag{5.14}$$

Writing the velocity in cylindrical coordinates we have

$$\frac{\partial}{\partial \mathbf{v}} = \widehat{\mathbf{v}}_\perp \frac{\partial}{\partial \mathrm{v}_\perp} + \widehat{\phi} \frac{1}{\mathrm{v}_\perp} \frac{\partial}{\partial \phi} + \widehat{\mathbf{e}}_{||} \frac{\partial}{\partial \mathrm{v}_{||}}$$

Where $\perp$ and $\|$ refer to the direction of $\mathbf{B}_0$. Since

$$\widehat{\varphi} = \widehat{\mathbf{e}} \times \widehat{\mathbf{v}}_\perp$$

We can rewrite (5.14) in the form

$$\mathbf{v} \cdot \nabla f_0 = \boldsymbol{\Omega}_c \frac{\partial f_0}{\partial \varphi} \tag{5.15}$$

showing that a gyrophase dependence of f_0 is associated with inhomogeneity (compare the dependence on the generalized moment in Chap. 4). We may also write (5.14) in the form

$$\mathbf{v} \cdot \left[\nabla f_0 + \Omega_c \left(\widehat{\mathbf{e}}_{||} \times \frac{\partial f_0}{\partial \mathbf{v}} \right) \right] = 0$$

Assuming $\mathbf{e}_{||}.\partial f_0 / \partial \mathbf{r} = 0$, we obtain the solution

$$\frac{\partial f_0}{\partial \mathbf{v}} = \frac{1}{\Omega_c} (\widehat{\mathbf{e}}_{||} \times \nabla f_0) + \widehat{\mathbf{v}}_{\perp} \frac{\partial f_0}{\partial \mathrm{v}_{\perp}} + \widehat{\mathbf{e}}_{||} \frac{\partial f_0}{\partial \mathrm{v}_{||}} \tag{5.16}$$

To first order we have

$$\frac{D_0 f_1}{Dt} = -\frac{q}{m} [\mathbf{E} + \mathbf{v} \times \delta \mathbf{B}] \cdot \frac{\partial f_0}{\partial \mathbf{v}} \tag{5.17}$$

where

$$\frac{D_0}{Dt} = \frac{\partial}{\partial t} + \mathbf{v} \cdot \nabla + \frac{q}{m} \mathbf{v} \times \mathbf{B}_0 \cdot \frac{\partial}{\partial \mathbf{v}}$$

is the operator along the unperturbed orbit. The unperturbed orbit is given by

$$\mathbf{v}(t) = \tilde{\mathbf{v}}(t) + \mathbf{v}_D(t) + \mathrm{v}_{||} \widehat{\mathbf{e}}_{||} \tag{5.18}$$

where $\tilde{\mathbf{v}}(t)$ is the pure gyromotion as given by (4.4), $\mathbf{v}_{\mathrm{D}}(\mathrm{t})$ is the magnetic drift which may be time dependent along the orbit and $\mathrm{v}_{\|}$ is the velocity along $\mathbf{B}_0$.

We now invert (5.17) as

$$f_1 = -\frac{q}{m} \int_{-\infty}^{t} [\mathbf{E}(\mathbf{r}(\mathrm{t}') + \mathbf{v}(t') \times \delta \mathbf{B}(\mathbf{r}(\mathrm{t}'))] \cdot \frac{\partial f_0}{\partial \mathbf{v}} dt' \tag{5.19}$$

Where $\mathbf{r}(\mathrm{t}')$ is the unperturbed orbit. Now, considering Fourier harmonics in time and space we obtain

$$f_k = -\frac{q}{m} \int_{0}^{\infty} [\mathbf{E}_{\mathrm{k}} + \mathbf{v} \times \mathbf{B}_{\mathrm{k}}] \cdot \frac{\partial f_0}{\partial \mathbf{v}} e^{-i\alpha(\tau)} d\tau \tag{5.20}$$

where

$$\alpha(\tau) = \mathbf{k} \cdot [\mathbf{r}(\mathrm{t}) - \mathbf{r}(\mathrm{t}')] - \omega\tau$$
$$= \frac{k_\perp \mathrm{v}_\perp}{\Omega_\mathrm{c}}[\sin(\Omega_c\tau + \varphi - \theta) - \sin(\varphi - \theta)] - \int_{t-\tau}^{t} \tilde{\omega}(t')dt'$$

and

$$\tilde{\omega} = \omega - k_{||}\mathrm{v}_{||} - \mathbf{k} \cdot \mathbf{v}_D(t), \tau = t - t' \quad \tilde{\mathrm{v}}(0) = \mathrm{v}_\perp(\cos\varphi, \sin\varphi)$$
$$\mathbf{k}_\perp = \mathrm{k}_\perp(\cos\theta, \quad \sin\theta)$$

We now introduce potentials i.e.

$$\mathbf{E} = -\nabla\varphi - \tfrac{\partial}{\partial \mathrm{t}}\mathbf{A} \quad \mathbf{B} = \nabla \times \mathbf{A}$$

For a Maxwellian distribution (5.16) now leads to

$$\mathbf{k} \cdot \frac{\partial f_0}{\partial \mathbf{v}} = i\frac{m}{T}(\mathbf{k}_\perp \cdot \hat{\mathbf{v}}_\perp + \mathrm{k}_{||}\mathrm{v}_{||} - \omega_{*f})f_0 \tag{5.21}$$

Where $\omega_{\bullet \mathrm{f}} = \mathrm{k} \bullet \mathbf{v}_{*\mathrm{f}}, \mathbf{v}_{*\mathrm{f}}, = (\mathrm{T}/(\mathrm{m}\Omega_\mathrm{c}))(\mathbf{e}_{||} \text{ x } \mathrm{grad}(\ln f_0).)$ As we have seen in Chaps. 3 and 4, $\mathrm{A}_{||}$ is the most important part of **A**. In order to include also compressional parts of the magnetic field perturbation i.e. $\delta\mathrm{B}_{||}$ (Eq. 6.18) we now include also an A_r component. This makes our choice of **A** general since we have the freedom of the gauge condition. We then find:

$$[\mathbf{E}_\mathrm{k} + \mathbf{v} \times \mathbf{B}_\mathrm{k}] \cdot \frac{\partial f_0}{\partial \mathbf{v}} = -i\frac{m}{T}f_0\mathbf{k}_\perp \cdot \mathbf{v}_\perp\varphi_k + i\frac{m}{T}(\omega - \omega_{*f})f_0\mathbf{A}_\mathbf{k} \cdot \mathbf{v}_\perp$$
$$- i\frac{m}{T}(k_{||}\varphi_k - \omega A_{k||})f_0\mathrm{v}_{||} + i\frac{m}{T}(\varphi_k - \mathrm{v}_{||}A_{k||})\omega_{*f}f_0 \tag{5.22}$$

Since now

$$\frac{d}{d\tau}e^{-i\alpha(\tau)} = i(\mathbf{k}_\perp \cdot \mathbf{v}_\perp(\tau) + \mathbf{k}_{||}\mathbf{v}_{||} - \omega)\mathbf{e}^{-\mathbf{i}\alpha(\tau)} \tag{5.23}$$

We may rewrite (5.20) in the form

$$f_k = \frac{q}{T}f_0\int_0^\infty \left[\varphi_\mathrm{k}\frac{d}{d\tau}e^{-i\alpha(\tau)} + i(\omega_{*f} - \omega)(\phi_k - \mathrm{v}_{||}A_{k||} - \mathbf{A}_\mathbf{k} \cdot \mathbf{v}_\perp)e^{-i\alpha(\tau)}\right] \cdot d\tau \tag{5.24}$$

or

$$f_k = \frac{q}{T}\varphi_\mathrm{k}f_0 + if_0\frac{q}{T}(\omega_{*f} - \omega)(\phi_k - \mathrm{v}_{||}A_{k||})\int_0^\infty e^{-i\alpha(\tau)}d\tau$$
$$- i\frac{q}{T}f_0(\omega_{*f} - \omega)\int_0^\infty \mathbf{A}_\mathbf{k} \cdot \mathbf{v}_\perp e^{-i\alpha(\tau)}d\tau \tag{5.25}$$

In an inhomogeneous system the orbit integrals in (5.25) require knowledge of an, in general, very complicate orbit. We will here avoid this complication by assuming the gyroperiod to be much shorter than any other time scale and performing an average over it. Thus a general orbit integral is written

$$\int_0^\infty G(\tau)d\tau = \sum_0^\infty \Delta\tau \frac{1}{\Delta\tau}\int_\tau^{\tau+\Delta\tau} G(\tau)d\tau \tag{5.26}$$

where $\Delta\tau$ is a gyroperiod and the integrals, normalized by $\Delta\tau$ are the local gyroaverages of an arbitrary function $G(\tau)$, subject only to the above assumption of time scales. In the gyroaveraging, we can ignore all variation on time scales longer than $\Delta\tau$. Since the time steps $\Delta\tau$ are small as compared to the longer time scales in the system we can convert the summation back to an integral over the long time scale. Thus

$$\int_0^\infty G(\tau)d\tau = \int_0^\infty <G(\tau)>d\tau$$

Now since

$$\exp\left[-i\frac{k_\perp v_\perp}{\Omega_c}\sin(\Omega_c\tau+\varphi-\theta)\right] = \sum_n J_n\left(\frac{k_\perp v_\perp}{\Omega_c}\right)\exp[-in(\Omega_c\tau+\varphi-\theta)]$$

We obtain

$$<e^{-i\alpha(\tau)}> = J_0(\xi)\exp[i\xi\sin(\varphi-\theta)]\exp\left[i\int_{t-\tau}^t \tilde{\omega}(t')dt'\right] \tag{5.27}$$

Where $\xi = k_\perp v_\perp/\Omega_c$.. Moreover, writing $\mathbf{A_k} \bullet \mathbf{v}_\perp = A_\perp v_\perp \cos(\Omega_c\tau+\varphi-\theta')$ we have

$$\begin{aligned}<\mathbf{A_k}\cdot\mathbf{v_k}e^{-i\alpha(\tau)}> &= A_\perp v_\perp <\cos(\Omega_c\tau+\varphi-\theta')e^{-i\alpha(\tau)}> \\ &= \frac{1}{2}e^{iL_k}.\left\langle\sum_n e^{in\theta}\{J_{n+1}\exp[-in(\Omega_c\tau+\phi)+i(\theta-\theta')]\right. \\ &\quad \left. + J_{n-1}\exp[in(\Omega_c\tau+\varphi)-i(\theta-\theta')]\}e^{+i\int_{t-\tau}^t \tilde{\omega}(t')dt'}\right\rangle \\ &= i\frac{v_\perp}{k_\perp}(\hat{\mathbf{e}}_\parallel \times \mathbf{k})\cdot\mathbf{A}\frac{dJ_0}{d\xi}e^{iL_k}e^{i\left[\int_{t-\tau}^t \tilde{\omega}(t')dt'\right]}\end{aligned}$$

where

$$L_k = \frac{k_\perp v_\perp}{\Omega_c}\sin(\varphi - \theta) = (\vec{v} \times \hat{e}_{||}) \cdot \vec{k}/\Omega_c$$

As mentioned above, the integration on the long timescale requires detailed knowledge of particle orbits. If we instead differentiate with respect to the long timescale we obtain the gyrokinetic equation

$$(\omega - k_{||\{}v_{||} - \omega_D)g_k = \frac{q}{T}(\omega - \omega_{*f})\left[(\phi_k - v_{||}A_{||})J_0(\xi) - i\frac{v_\perp}{k_\perp}(\hat{\mathbf{e}}_{||} \times \mathbf{k}) \cdot \mathbf{A_k}J_0'\right]f_0 \tag{5.28}$$

Where

$$g_k = \left(f_k + \frac{q\phi}{T}f_0\right)e^{-iL_k}$$

$$J_0' = \frac{dJ_0}{d\xi}$$

$$\xi = \frac{k_\perp v_\perp}{\Omega_c}$$

$$L_k = (\mathbf{v} \times \hat{\mathbf{e}}_{||}) \cdot \frac{\mathbf{k}}{\Omega_c}$$

Here $\omega_D = \mathbf{k} \bullet \mathbf{v}_D(v_{||}^2, v_\perp^2)$ as given in the derivation of the drift kinetic equation. The diamagnetic drift frequency contains grad f_0 and is also velocity dependent. For a Maxwellian distribution where both n and T are space dependent we find

$$\omega_{*f} = \omega_*\left[1 + \eta\left(\frac{mv^2}{2T} - \frac{3}{2}\right)\right] \tag{5.29}$$

Where $\eta = L_n/L_T$ and ω_* is the usual fluid diamagnetic drift with only a density gradient. Equation 5.28 agrees with the gyrokinetic equation obtained by Antonsen and Lane [6.53].

5.2.1 Applications

It is straightforward to rederive the results on both electrostatic and electromagnetic modes in Chap. 4 from (5.28). The advantage of (5.28) is that it allows for space dependent coefficients (i.e. $\omega_{\bullet}$, ω_D etc.). We also note that $k_{\|}$ in general has to be treated as an operator. Another major difference is that (5.28) is valid for arbitrary ω_D/ω while the treatment in Chap. 4 only works to first order in ω_D/ω. We will here explore this property a little in the electrostatic limit. We will also take the limit $\omega >> k_{\|} v_{th}$ in which case $k_{\|}$ can be omitted. In this case the density response may be written:

$$\frac{\delta n}{n} = -\frac{q\phi}{T}\left[1 - \frac{1}{n_0}\int \frac{\omega - \omega_{\bullet}[1 + \eta(mv^2/2T - 3/2)]J_0{}^2(\xi_k) f_M}{\omega - \omega_D\left(v_{\|}{}^2 + v_{\perp}{}^2/2\right)/v^2{}_{th}} d^3v\right] \tag{5.30}$$

where ω_D is the fluid magnetic drift frequency and all velocity dependence has been written explicitly. For comparison with fluid theory it is useful to expand (5.30) for $\omega_D/\omega \ll 1$ and $\xi = k\rho \ll 1$. Including terms up to second order in both small parameters we have

$$\frac{\delta n}{n} = \Big[\frac{\omega_{*e}}{\omega} - \left(1 - \frac{\omega_{*i}}{\omega}(1+\eta_i)\right)\left(1 + \Gamma\frac{\omega_{Di}}{\omega}\right)\left(\frac{\omega_{De}}{\omega} + k_{\perp}{}^2\rho^2\right)$$
$$+ \Gamma\eta_i \frac{\omega_{*i}\omega_{Di}\omega_{De}}{\omega^3}\Big]\frac{e\phi}{T_e} \tag{5.31}$$

Where $\omega_{\bullet iT} = \omega_{\bullet}(1+\eta_i)$ and $\Gamma = 7/4$. The expansion in ω_D has a very limited regime if applicability in tokamaks. However this is really the only way of comparing with advanced fluid theory analytically. As we will see in Chap. 6 the same expansion, except that $\Gamma = 5/3$ (5% difference) is obtained for a reactive fluid closure including the diamagnetic heatflow.

We note that (5.31) is also useful for MHD modes since for these ions can usually be treated in the electrostatic limit. For these modes the natural linear eigenfrequency is $\omega_{\bullet iT}$ at which the second part of (5.31) vanishes. The last term here acts as an additional driving pressure force which is responsible for an instability below the MHD beta limit (Eq. 6.70). As can be seen from the advanced fluid model presented later the last term in (5.31) is due to the divergence of the diamagnetic heat flow which is the term in the energy equation that corresponds to the lowest order driving term in the continuity equation, i.e. the divergence of the diamagnetic particle flux.

Another property of (5.31) is that to first order in $\omega_{\bullet}/\omega$ and ω_D/ω it reduces to $\omega_{\bullet}(1-\varepsilon_n)/\omega$ where $\varepsilon_n = \omega_D/\omega_{\bullet}$. Since $\omega_{\bullet}/\omega$ leads to the main driving term for interchange and ballooning modes in MHD, the ε_n part is the main reason for the reduction of the growthrate of MHD ballooning modes for large ε_n seen in kinetic theory (6.70).

However, the most interesting aspect of (5.31) is probably that the last term is the first in such an expansion to separate between temperature and pressure gradients. This is so since the first term on the right hand side is due to $E \times B$ convection and thus will cancel with the corresponding electron term when we derive a dispersion relation from the condition div $\mathbf{j} = 0$ and in the other terms density and temperature gradients enter together to form a pressure gradient. Then the last term, as pointed out above, comes from the heat flow. Thus an adiabatic model would not separate out temperature gradients (degenerate case).

5.3 The Nonlinear Gyrokinetic Equation

We will now continue the above derivation of a gyrokinetic equation to the nonlinear regime.

It is straight forward to continue the linear derivation iteratively, thus inserting the linear relations in the nonlinear terms. This corresponds to an expansion in the perturbations but this is, in fact, allowed since in the main part of a tokamak, $e\varphi/T \sim 10^{-2}$. Moreover, the Hasegawa-Mima equation, which is regarded as fully nonlinear, emerges in the appropriate limit. We will here, for simplicity, omit $\mathbf{A}\perp$.

A convenient way of writing (5.28) is then:

$$f_k{}^{(1)} = -\frac{q}{T} f_0 H_k \tag{5.32a}$$

$$H_k = \phi_k + \Gamma_k e^{iL_k(\mathbf{v})} \tag{5.32b}$$

$$\Gamma_k = \frac{\omega_* - \omega}{\tilde{\omega}} \left[\phi_k + v_{||} A_{||}\right] J_0(\xi_k) \tag{5.32c}$$

$$L_k = (\mathbf{v} \times \mathbf{e}_{||}) \cdot \mathbf{k} / \mathbf{\Omega_c} \tag{5.32d}$$

For the background variation we will use the formulation

$$\frac{\partial f_0}{\partial \mathbf{v}} = -\left(\frac{\kappa}{\Omega_c} \hat{\boldsymbol{\theta}} + \frac{m}{T} \mathbf{v}\right) f_0(r, \mathbf{v}) \tag{5.33}$$

An interesting point is that linearly the phase dependence of e^{iL_k} disappears into a Bessel function while it plays an important role in the integration of the nonlinear terms. The nonlinearity we are interested in can be written

$$f_{k-k'}{}^{(2)}(t') = -\frac{q}{m} \int_0^\infty \left[\mathbf{E}_{\mathbf{k}''} + (\mathbf{v} \times \mathbf{B}_{\mathbf{k}''})\right] \cdot \frac{\partial f^{(1)}{}_{k-k'-k''}(\tau')}{\partial \mathbf{v}} e^{-i\alpha_{k-k'}(\tau')} d\tau' \ldots; \qquad (\tau' = t' - t'') \tag{5.34}$$

Here we need to calculate

$$\frac{\partial f_k{}^{(1)}}{\partial \mathbf{v}} = \frac{q}{T}\left(\frac{\kappa}{\Omega_c}\widehat{\theta} + \frac{m}{T}\mathbf{v}\right) f_0 H_k - \frac{q}{T} f_0 \left[\frac{\partial H_k}{\partial \mathrm{v}_\perp}\widehat{\mathbf{v}}_\perp + \frac{1}{\mathrm{v}_\perp}\frac{\partial H_k}{\partial \varphi}\widehat{\varphi} + \frac{\partial H_k}{\partial \mathrm{v}_{||}}\widehat{e}_{||}\right] \quad (5.35)$$

Which we substitute into (5.34) as

$$\begin{aligned}
f_{k,\omega}{}^{(2)} &= \left(\frac{q}{m}\right)^2 f_0(r,\mathbf{v}) \\
&\times \sum_{\mathbf{k}'} \int_0^\infty \left\{ i\mathbf{k}' \cdot \mathbf{v} \begin{bmatrix} \phi_{\mathbf{k}'} H_{k-k'} - \dfrac{T}{m\mathrm{v}_\perp}\left(\phi_{k'} - \mathrm{v}_{||} A_{||k'}\right)\dfrac{\partial H_{k-k'}}{\partial \mathrm{v}_\perp}. \\ \\ + \dfrac{T}{m} A_{||k'} \dfrac{\partial H_{k-k'}}{\partial \mathrm{v}_{||}} \end{bmatrix} \right\} e^{-i\alpha_{k-k'}(\tau)} d\tau \\
&- \left(\frac{q}{m}\right)^2 f_0(r,\mathbf{v}) \sum_{\mathbf{k}'} \int_0^\infty \left\{ i\mathbf{k}' \cdot \left(\widehat{\mathbf{e}}_{||} \times \widehat{\mathbf{v}}_\perp\right) \frac{T}{m\mathrm{v}_\perp}\left(\phi_{k'} - \mathrm{v}_{||} A_{||k'}\right) \frac{1}{\mathrm{v}_\perp}\frac{\partial H_{k-k'}}{\partial \varphi} \right\} e^{-i\alpha_{k-k'}(\tau)} d\tau \\
&- \left(\frac{q}{m}\right)^2 f_0(r,\mathbf{v}) \sum_{\mathbf{k}'} \int_0^\infty \left\{ \begin{array}{c} \left[\mathrm{v}_{||}\left(k_{||}'\phi_{k'} - \omega' A_{||k'}\right) - \omega_*'\left(\phi_{k'} - \mathrm{v}_{||} A_{||}\right)\right] \\ \\ \times H_{k-k'} + i\dfrac{T}{m}\left(k_{||}'\phi_{k'} + \omega' A_{||k'}\right) \dfrac{\partial H_{k-k'}}{\partial \mathrm{v}_{||}} \end{array} \right\} e^{-i\alpha_{k-k'}(\tau)} d\tau
\end{aligned} \quad (5.36)$$

Here we assume ω' to be associated with k' and ω'' with $\mathrm{k}'' = \mathrm{k} - \mathrm{k}'$.

Rewriting (5.36) in a similar way as (5.25) the following integrals appear:

$$G(\xi) = i\int_0^\infty \mathbf{k}' \cdot \mathbf{v} e^{-i\alpha_k(\tau)} d\tau \quad (5.37a)$$

$$R(\xi') = i\int_0^\infty \mathbf{k}' \cdot \mathbf{v} \frac{\partial L_{k-k'}}{\partial \mathrm{v}_\perp} e^{-i\Theta(\tau)} d\tau \quad (5.37b)$$

$$Q(\xi') = i\int_0^\infty \mathbf{k}' \cdot \left(\widehat{\mathbf{e}}_{||} \times \widehat{\mathbf{v}}_\perp\right) \frac{\partial L_{k-k'}}{\partial \varphi} e^{-i\Theta(\tau)} d\tau \quad (5.37c)$$

Where

$$\Theta(\tau) = \alpha_k(\tau) - L_{k-k'}(\tau)$$

Using (5.37a–c) we can rewrite (5.36) in the much simpler form

$$f^{(2)}{}_{k,\omega} = \left(\frac{q}{T}\right)^2 f_0 \sum_{k'} \left\{ \phi_{k'}\phi_{k-k'}G(\xi) - \frac{T}{m\mathrm{v}_\perp}\left[\phi_{k'} - \mathrm{v}_{||}A_{||}\right] i\Gamma_{k-k'}(R(\xi') + Q(\xi')) \right\} \tag{5.38}$$

The integration of the expressions (5.37) proceed in a way very similar to the evaluation of the $\mathbf{A}_\perp$ part in the linear gyrokinetic equation (Eqs. 5.27–5.28) with the result:

$$G(\xi) = -i\frac{\mathrm{v}_\perp}{\mathrm{k}_\perp\tilde{\omega}}(\mathbf{k}\times\mathbf{k}')\cdot\widehat{\mathbf{e}}_{||}J_0{}'(\xi)e^{iL_k} \tag{5.39a}$$

$$R(\xi') + Q(\xi') = \frac{\mathrm{v}_\perp}{\tilde{\omega}\Omega_c}(\mathbf{k}'\times\mathbf{k}'')\cdot\widehat{\mathbf{e}}_{||}J_0(\xi')e^{iL_k} \tag{5.39b}$$

Where we introduced $\mathbf{k}'' = \mathbf{k} - \mathbf{k}'$ and $\mathrm{J}_0{}' = \mathrm{dJ}_0/\mathrm{d}\xi$.

The new feature of the nonlinear terms is that the vector products appear. They are initially obtained as terms of the type proportional to e.g. $\sin(\theta-\theta')$ etc. Thus it is essential to have $\mathbf{k}'$ different from $\mathbf{k}$ in (5.37). We now observe that $\mathrm{G}(\xi)$ vanishes upon summation over $\mathbf{k}'$. This corresponds to terms prop to $\mathbf{v}_\mathrm{E}\bullet\mathrm{grad}\ \varphi$. We then arrive at

$$f^{(2)}{}_{k,\omega} = \frac{q^2}{T}f_0\frac{i}{m\tilde{\omega}\Omega_c}\sum_{k'k''}(\mathbf{k}'\times\mathbf{k}'')\cdot\widehat{\mathbf{e}}_{||}e^{iL_k}\left(\phi_{k'} - \mathrm{v}_{||}A_{||k'}\right)\left(\phi_{k''} - \mathrm{v}_{||}A_{||k''}\right)$$
$$\times\frac{\omega_*{}'' - \omega''}{\tilde{\omega}''}J_0(\xi')J_0(\xi'') \tag{5.40}$$

We now obtain a nonlinear dispersion relation of the form:

$$D(\omega,\mathbf{k})\phi_{\mathbf{k},\omega} = \frac{T_i}{e}\frac{\omega^2}{\Lambda_0 - 1}\left(\frac{\delta n^{(2)}{}_{e,k,\omega}}{n_0} - \frac{\delta n^{(2)}{}_{i,k,\omega}}{n_0}\right) \tag{5.41}$$

As an example we will now consider the electrostatic approximation. We will furthermore ignore magnetic drifts. We then obtain the equation:

$$\left\{\omega\left[1 + \frac{T_e}{T_i}(1 - \Lambda_0(s)\right] - \omega_{*e}\Lambda_0(s)\right\}\phi_{\mathbf{k},\omega}$$
$$= -\frac{1}{B_0 n_0}\frac{T_e}{T_i}\int f_0\sum_{k'}(\mathbf{k}'\times\mathbf{k}'')\cdot\widehat{\mathbf{e}}_{||}\left(J_0{}^2(\xi') - 1\right)\left(1 - \frac{\omega_*{}''}{\omega''}\right)J_0{}^2(\xi'')$$
$$\times\,\phi_{k'}\phi_{k''}d\mathbf{v} \tag{5.42}$$

We now take the limit $k^2\rho^2 \ll 1$ keeping only first order terms and limit consideration to three waves. We can then perform integration over velocity space obtaining:

$$\left\{\omega\left(1+\frac{1}{2}k^2\rho^2{}_s\right) - \omega_{*e}\left(1-\frac{1}{2}k^2\rho_i{}^2\right)\right\}\phi_{\mathbf{k},\omega}$$
$$= -\frac{\rho_s{}^2}{B_0}\frac{T_e}{T_i}(\mathbf{k}'\times\mathbf{k}'')\cdot\widehat{\mathbf{e}}_{\|}\left[k_\perp{}'^2\left(1-\frac{\omega''{}_{*i}}{\omega''}\right) - k_\perp{}''^2\left(1-\frac{\omega_{*i}{}'}{\omega'}\right)\right]\phi_{k'}\phi_{k''} \quad (5.43)$$

This is the Hasegawa Mima equation in (ω,k) space generalized to include first order FLR effects. We note that since this is a nonlinear equation, ω here includes nonlinear frequency shifts. We note that the role of $k_\perp{}^2$ in the cascade rules is now replaced by:

$$k_\perp{}'^2 - k_\perp{}''^2 \rightarrow \left[k_\perp{}'^2\left(1-\frac{\omega''{}_{*i}}{\omega''}\right) - k_\perp{}''^2\left(1-\frac{\omega_{*i}{}'}{\omega'}\right)\right] \quad (5.44)$$

We note, however that here (ω,**k**) have been kept together. In a strongly nonlinear situation we should perform the summations over ω and **k** independently.

5.4 Gyro-Fluid Equations

Since we have here derived kinetic equation fluid that have been averaged over the gyromotion it may be useful to discuss briefly fluid-type equations obtained by taking moments of these averaged kinetic equations. We have, actually already derived such an equation in (5.11). However, this equation does not contain higher order FLR effects. A special feature of gyro-fluid equations is that they contain only guiding centre drifts, i.e. there are no diamagnetic or stress tensor drifts. Still they contain full FLR effects if derived from the gyrokinetic equation. Of course gyrofluid equations contain less information than the full fluid equations since they have been averaged over the gyromotion. They are, however equivalent to the fluid equations obtained by the low frequency expansion in Chap. 2. In order to obtain higher order FLR effects from the fluid equations we, however, have to make use of the stress tensor as shown in Chap. 2. This is often tedious. Since the perpendicular dynamics is due to the guiding centre drifts which we already know, the main remaining questions now concern the parallel motion. It appears that the first gyro-fluid equation for parallel motion, including magnetic drifts was obtained by Waltz et al. [18] (5.45). We note that this equation has a convective magnetic drift included. This may be surprising since the parallel motion should be the same in gyrofluid and fluid equations and magnetic drifts are not present in fluid equations. This was resolved in [19] when (5.45) was rederived from fluid theory using the stress tensor with magnetic curvature effects. Thus the perpendicular and parallel dynamics are coupled by magnetic curvature.

$$\frac{\partial \delta \mathrm{u}_{||}}{\partial \mathrm{t}} + 2\mathbf{v_D} \cdot \nabla \delta \mathrm{u}_{||} = -\widehat{e}_{||} \cdot \nabla(\delta p + en\phi) \tag{5.45}$$

A general formalism including magnetic drifts in the stress tensor was presented in [20].

References

1. P.H. Rutherford and E.A. Frieman, Phys. Fluids **11**, 569 (1968).
2. C.S. Liu, Phys. Fluids **12**, 1489 (1969).
3. J.C. Adam, W.M. Tang and P.H. Rutherford, Phys. Fluids **19**, 1561 (1976).
4. B. Coppi and F. Pegoraro, Nuclear Fusion **17**, 969 (1977).
5. A. Hasagawa and K. Mima, Phys. Fluids **21**, 87 (1978).
6. W.M. Manheimer and T.M. Antonsen Jr, Phys. Fluids **22**, 957 (1979).
7. T.M. Antonsen and B. Lane, Phys. Fluids **23**, 1205 (1980).
8. C.Z. Cheng, Phys. Fluids **25**, 1020 (1982).
9. E.A. Frieman and L. Chen Phys. Fluids **25**, 502, (1982)
10. P.N. Guzdar, Liu Chen, W.M. Tang and P.H. Rutherford, Phys. Fluids **26**, 673 (1983).
11. J.Weiland, Physica Scripta **29**, 234 (1984).
12. W.M. Tang, G. Rewoldt and L. Chen, Phys. Fluids **29**, 3715 (1986).
13. R.R. Dominguez and R.R. More, Nuclear Fusion **26**, 85 (1986).
14. M. Liljeström and J. Weiland, Phys. Fluids **31**, 2228 (1988).
15. T.S. Hahm, Phys. Fluids **31**,2670, (1988).
16. M. Liljeström and J. Weiland, Phys. Fluids **31**, 2228 (1988).
17. J. Nilsson, M. Liljeström and J. Weiland, Phys. Fluids **B2**, 2568 (1990).
18. R.E. Waltz, R.R. Dominguez and G.W. Hammett Phys. Fluids **B4**, 3138 (1992).
19. D. Strinzi, A.G. Peeters and J. Weiland, Phys. Plasmas **15**, 044502 (2008).
20. J.J. Ramos, Phys. Plasmas **12**, 112301 (2005)

Chapter 6
Low Frequency Modes in Inhomogeneous Magnetic Fields

We have now seen how some typical low frequency modes can be driven unstable by density, pressure or current gradients in simple geometries. A more accurate description of collective modes in magnetic confinement systems, in general, requires more detailed geometry effects as well as separate effects of density and temperature gradients [1–197]. In the present chapter we will aim at making the geometrical description more accurate, thus in most cases leading to eigenvalue problems for the modes concerned. We will also derive a more complete drift kinetic description, introduce the gyrokinetic equation and present an advanced fluid model. We will furthermore review briefly the fields of transport due to magnetic fluctuations and advanced fluid models.

6.1 Anomalous Transport in Systems with Inhomogeneous Magnetic Fields

Although work on understanding transport in magnetic confinement systems has been going on for about 60 years, this problem is still an ultimate scientific issue [167]. Its importance for the size and cost of a reactor is obvious and critical but the scientific difficulties associated with it are enormous.

Initially, for Ohmically heated plasmas, the interest was mainly focused on electron transport since it dominated. While the particle transport has to be ambipolar, the energy transport does not. Thus electron thermal transport, through magnetic perturbations is an obvious option. The most well known scaling law in this regime is the Alcator scaling [31]:

$$\tau_E \propto na^2 \tag{6.1}$$

J. Weiland, *Stability and Transport in Magnetic Confinement Systems*, Springer Series on Atomic, Optical, and Plasma Physics 71,
DOI 10.1007/978-1-4614-3743-7_6,

Several theories have been able to recover the density dependence through a dependence on collisionality. A candidate for magnetic transport is the microtearing mode [57, 58] while the dissipative trapped electron mode [10] can give such a scaling through electrostatic dynamics [85]. We shall return later to trapped electron modes and will here discuss magnetic perturbations briefly. Several papers consider the transport for given magnetic fluctuations (or islands). Critical for transport is the island width which determines whether islands from neighbouring rational surfaces overlap. When this is the case we can use the Rechester-Rosenbluth diffusion coefficient [41].

$$D = \mathrm{v}_{\mathrm{the}} L_c \left(\frac{\delta B}{B} \right) \tag{6.2}$$

where L_c is the correlation length which in general, through the mode width depends on the resistivity. An obvious candidate for creating magnetic perturbations is the magnetic drift mode (4.59). This mode has $k_\| = 0$ and $\varphi = 0$. In a realistic plasma with magnetic shear, this mode is localized near rational surfaces (see the following section), where $k_\| \approx 0$. In such a geometry we have a radial eigenvalue problem with the boundary conditions $\varphi = 0$ (odd φ) and even $A_\|$ at the rational surface. Such a mode is a tearing mode [4] which is destabilized by resistivity.

While most confinement systems are designed so as to eliminate the dangerous global tearing modes with $k_\theta L_n \sim 1$, a localized "micro tearing" mode with $k_\theta L_n >> 1$ can still be unstable if $\nu_{ei} > \omega_{*e}$. The saturation level, due to diffusion, is [57]:

$$\frac{\delta B}{B} = \frac{\rho_e}{L_{T_e}} \tag{6.3}$$

This level is of the order 10^{-5} to 10^{-4} in typical tokamaks. This mode is thus a candidate for explaining the Alcator scaling in the Ohmic regime which usually is collision dominated. It is, however, almost stable in collisionless plasmas, giving a very small transport.

In the collisionless regime, electromagnetic drift wave turbulence has been considered as a candidate for generating magnetic transport [34, 88, 97]. The magnetic fluctuation level is, however, usually too low or the correlation length too short due to very high mode numbers. A remaining possibility is nonlinearly self-sustained magnetic perturbations. As an example collisionless tearing modes can be driven unstable by the turbulent radial diffusion of electrons [119]. The experimental situation remains unclear. On the one hand evaluations of the magnetic flutter transport on TEXT [125] conclude that it is considerably smaller than the total transport while experiments on Tore Supra [126] indicate the presence of magnetic islands. A recent development in this field is the current diffusive ballooning mode [124]. It is a MHD type mode which is described by resistive MHD equations.

A transport model has been based on this mode. It tends to give good agreement with experiments for electron thermal diffusion but not so good agreement for ion diffusion [182]. This may be due to the fact that only one fluid equations are used and a full kinetic derivation is still lacking.

When the density is increased sufficiently, the confinement time saturates and another instability takes over. Transport code simulations [86, 87] indicate that this is the ion temperature gradient driven mode [1, 28, 61].

6.2 Toroidal Mode Structure

A general plasma perturbation in a torus must in order to fulfil the boundary conditions be a superposition of elementary perturbations of the form

$$f(r,\theta,\phi) = \widehat{f}(r)e^{i(m\theta - n\phi)} \tag{6.4}$$

where θ is the poloidal and φ is the toroidal angle according to Fig. 6.1. The phase angle can be represented as

$$m\theta - n\phi = k_r\, r\theta + k_\phi R\phi$$

Where

$$k_r = \frac{m}{r} \qquad \text{and} \qquad k_\phi = -\frac{n}{R}$$

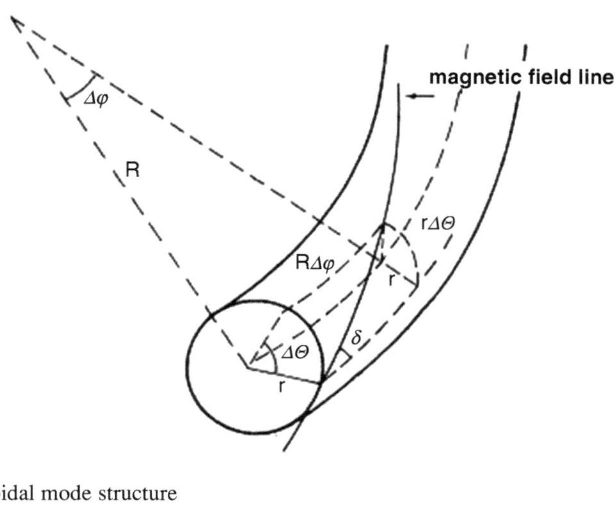

Fig. 6.1 Toroidal mode structure

We here allow both positive and negative mode number corresponding to propagation in different directions. Thus the choice of minus sign for k_φ is arbitrary and was chosen in order to have the possibility of cancellation between the poloidal and toroidal phases for positive modenumbers. Such a cancellation is of special interest since there the parallel k vanishes, corresponding to the most unstable situation as discussed earlier.

The magnetic field can be written:

$$\mathbf{B} = \mathrm{B}_\theta \hat{\mathbf{e}}_\theta + \mathrm{B}_\phi \hat{\mathbf{e}}_\phi$$

Thus

$$\mathbf{k} \cdot \mathbf{B} = \frac{m}{r} \mathrm{B}_\theta - \frac{n}{R} \mathrm{B}_\phi$$

where, in a tokamak $\mathrm{B}_\varphi >> \mathrm{B}_\theta$, $R >> r$. Thus the two contributions to **k•B** are usually comparable. It is therefore convenient to introduce

$$q(r) = \frac{\Delta\phi}{\Delta\theta} = \frac{B_\phi}{B_\theta} \frac{r}{R} \tag{6.5}$$

where $\Delta\varphi$ and $\Delta\theta$ are the changes in φ and θ when we follow a field line as shown in Fig. 6.1. Here q is called the rotational transform which is one of the key parameters for tokamak stability. It is thus also called safety factor. In terms of q we can then write

$$\mathbf{k} \cdot \mathbf{B} = \frac{n}{r} \mathrm{B}_\theta \left[\frac{m}{n} - q(r) \right] \tag{6.6}$$

showing that $\mathbf{k} \bullet \mathbf{B} = \mathbf{0}$ when $q(r) = m/n$. This means that the pitch angle of an equiphase line $\alpha = (r/R)m/n$ coincides with the pitch angle of the magnetic field lines $\delta = r\Delta\theta/(R\Delta\varphi)$. In this situation $k_\| = 0$ and the electrons cannot cancel space charge caused by the mode (m,n) on the magnetic surface (surface containing magnetic field lines) corresponding to $q(r) = m/n$. This surface is called the rational surface. Since q(r) usually is growing monotonously with r, each mode will not have more than one rational surface. Modes that are well localised around the rational surface are usually more unstable since the effective $k_\|$ is small. One common way of expressing $k_\|$ is by rewriting (6.6) as

$$k_\| = \frac{\mathbf{k} \cdot \mathbf{B}}{B} = \frac{n}{r} \frac{\mathrm{B}_\theta}{\mathrm{B}_\phi} \left[\frac{m}{n} - q(r) \right] = [m - nq(r)]/Rq \tag{6.7}$$

Where we assumed that $\mathrm{B}_\theta << \mathrm{B}_\varphi$ so that $B \approx \mathrm{B}_\varphi$. The wave number 1/Rq represents the inhomogeneity of the magnetic field and is related to the connection length L_c defined by

$$L_c = 2\pi Rq$$

L_c is a measure of the length of a magnetic field line between two points with the same θ. We then introduce

$$k_c = \frac{2\pi}{L_c} = \frac{1}{qR}$$

So that

$$k_{||} = [m - nq(r)]k_c \tag{6.8}$$

Now Taylor expanding q(r) around the rational surface, r_0 defined by $q(r_0) = m/n$ we obtain

$$k_{||} = -nk_c \frac{dq}{dr}(r - r_0) = -\frac{r - r_0}{L_s} k_\theta \tag{6.9}$$

Where we introduced the shear length

$$L_s = \frac{q}{k_c} \frac{1}{rdq/dr} \tag{6.10}$$

A frequently used measure of the shear strength is also

$$s = \frac{d\ln q}{d\ln r} = \frac{r}{q}\frac{dq}{dr} \tag{6.11}$$

These two parameters are related through

$$L_s = \frac{Rq}{s} = (sk_c)^{-1}$$

For a tokamak, typically, s is small near the axis and is otherwise of order 1. Another quantity which is often of interest is the distance between neighbouring rational surfaces. If $q(r_0) = m/n$ and $q(r_0 + \Delta r) = (m + 1)/n$ we obtain for slowly varying q(r)

$$\Delta r = \left(n\frac{dq}{dr}\right)^{-1}$$

If we instead vary n, an additional factor q will appear. Since a mode usually is localised around its rational surface, the question of overlapping between two modes and accordingly nonlinear interaction and transport properties, depends strongly on the distance between rational surfaces.

Another inhomogeneity which is important for the modestructure is the decrease of B_φ along the main radius. This variation can be expressed as

$$B_\phi = \frac{B_T}{1 + (r/R)\cos\theta} \tag{6.12}$$

Modes which are driven by the curvature of the magnetic field lines are usually strongly influenced by the different sign of the curvature on the outside and inside of the torus, introducing a periodicity with period L_c along the magnetic field lines. Although it is still possible to Fourier decompose the modes into components of the type (6.4) these components will be linearly coupled and an eigenmode will now have the form

$$f(r,\theta,\phi) = \widehat{f}(r,\theta)e^{i(m\theta - n\phi)} \tag{6.13}$$

This leads to a two dimensional problem for the mode structure which in general is difficult to treat exactly and approximate analytical solutions are usually only available if the r or θ dependence dominates.

The poloidal variation of f is often, by projection, transferred to a variation along the magnetic field. A Fourier-decomposition along the magnetic field then leads to a coupling between components with different $k_{\|}$. Since a convection in the radial direction changes $k_{\|}$ we realize that we will obtain a coupling between the mode structure along the magnetic field and the position in the radial direction. This coupling usually tends to inhibit the radial convection, thus reducing the shear damping. Since $f(r,\theta)$ will vary at least as fast as the fundamental mode $m = 1$ it is not necessary to distinguish between the modes m and $m + 1$. It is instead common to express an eigenmode by its independent mode number, n. Close to the rational surface the poloidal variation is described by $m = q(r)n$ and an additional variation of f. Due to the poloidal variation we must, however, also introduce a θ dependent safety factor $\nu(r,\theta)$ so that

$$q(r) = \frac{1}{2\pi}\oint \nu(r,\theta)d\theta$$

The representation (6.13) then turns into

$$f(r,\theta,\phi) = \widehat{f}(r,\theta)e^{in\left[\int^{\theta}(\nu(r,\theta)d\theta - \phi)\right]} \tag{6.14}$$

This is a very useful eikonal description which for large n describes a mode with a rapid variation across the magnetic field and a slow variation along the magnetic field. The r dependent helicity that results from putting $m = nq(r)$ corresponds to a mode that tends to follow the field lines when it moves in the radial direction, thereby minimising $k_{\|}$ and the restoring line bending force. There is, however, one disadvantage which has been discussed extensively in connection with electromagnetic ballooning modes. It is the lack of periodicity of the phase function at a distance from the rational surface. This cannot be compensated by the amplitude f within the eikonal description. The problem was solved by transforming the problem to an infinite domain in θ where no periodicity is required and then constructing a periodic solution by adding the integer Fourier components [26, 39, 42].

6.3 Curvature Relations

We will now discuss some fundamental relations obtained in a curved magnetic field from a fluid point of view. We start by noting that a curved magnetic field always also has a gradient perpendicular to the field due to the fact that the magnetic field is divergence free. In order to simplify we will often just use curvature although we understand that a curvature drift also means the presence of a gradient B drift and these are combined to the magnetic drift. We start with the condition for pressure balance. By adding the equations of motion for ions and electrons and dropping the inertial terms (error of order ω/Ω_c) we obtain

$$\nabla p = \mathbf{j} \times \mathbf{B} \tag{6.15}$$

Combining (6.15) with

$$\nabla \times \mathbf{B} = \mu_0 \mathbf{j} \tag{6.16}$$

We obtain the pressure balance equation

$$\nabla\left(p + \frac{B^2}{2\mu_0}\right) = \frac{1}{\mu_0}(\mathbf{B} \cdot \nabla)\mathbf{B} \tag{6.17}$$

where $(1/2\mu_0)B^2$ is the magnetic field pressure and $(\mathbf{B}\bullet\text{grad})\mathbf{B}$ is the field curvature. When written for the background quantities (6.17) shows how the magnetic field pressure varies in space due to particle pressure (diamagnetic effect) and field curvature. In a low β plasma the pressure gradient term is often neglected and (6.17) then just gives the geometrical relation of the vacuum field. If we, on the other hand, write (6.17) for perturbed quantities and linearise we observe that

$$(\mathbf{B} \cdot \nabla)\mathbf{B} \approx \mathbf{B_0} k_{\|}\, \delta\mathbf{B} + (\delta\mathbf{B} \cdot \nabla)\mathbf{B_0} \approx \frac{1}{R}\delta B \mathbf{B_0}$$

for $k_{\|} \sim 1/R$ where R is the radius of curvature of the background field. This estimate is typical for quasi-flute modes in toroidal machines and since $\text{grad}(B)^2 \sim k_\perp B_0 \delta B$ we realize that the curvature term is normally negligible for perturbations. We then have

$$\nabla\left(\delta p+\frac{\delta B^2}{2\mu_0}\right)=0$$

Since $\delta \mathrm{B}^2=(\mathbf{B_0}+\delta\mathbf{B})^2\approx 2\mathbf{B}_0\bullet\delta\mathbf{B}$ we find the relation

$$\delta B_{||}=-\frac{\mu_0\,\delta p}{B_0} \tag{6.18}$$

Which relates the parallel perturbation in B to the pressure perturbation. This can be seen as a consequence of the magnetic confinement and the pressure balance. We now return to the derivation of the drift velocities in Chap. 2. Introducing $\hat{e}_{||}=\mathbf{B}_0/\mathrm{B}$ we have

$$\widehat{\mathbf{e}}_{||}\times(\mathbf{v}\times\mathbf{B})=\mathbf{v}\mathrm{B}_{||}-\mathbf{B}\mathrm{v}_{||}=\mathbf{v}_{\perp}\mathrm{B}_0\left(1+\frac{\delta\mathrm{B}_{||}}{\mathrm{B}_0}\right)$$

Then linearising the expression for $\mathbf{v}_\perp$ we find, dropping $\mathbf{v}_\pi$ and $\mathbf{v_g}$ that the only drift which is modified is $\mathbf{v}_{*}$. The quantity usually needed in the derivation of dispersion relations is div$\mathbf{j}$. We here note that it is the total parallel magnetic field (including perturbation) that should appear in the denominator of the fluid drifts. Thus we are interested in evaluating the expression

$$\nabla\cdot\left[(n_i\mathbf{v}_{*\mathrm{i}}-n_e\mathbf{v}_{*\mathrm{e}})\left(1-\frac{\delta\mathrm{B}}{\mathrm{B}_0}\right)\right]\approx\nabla\cdot\left[\frac{1}{\mathrm{eB}_0}(\widehat{\mathbf{e}}_{||}\times\nabla p)\left(1-\frac{\delta\mathrm{B}_{||}}{\mathrm{B}_0}\right)\right]\approx$$

$$\approx-\frac{1}{\mathrm{eB}_0}\frac{\nabla\mathrm{B}_0}{\mathrm{B}_0}\cdot(\widehat{\mathbf{e}}_{||}\times\nabla\delta p)+\frac{1}{\mathrm{eB}_0}\nabla\cdot(\widehat{\mathbf{e}}_{||}\times\nabla\delta p)-\frac{1}{\mathrm{eB}_0}(\widehat{\mathbf{e}}_{||}\times\nabla\delta p)\cdot\frac{\nabla\delta\mathrm{B}_{||}}{\mathrm{B}_0}$$

$$=\frac{1}{\mathrm{eB}_0}\left(\widehat{\mathbf{e}}_{||}\times\frac{\delta B_{||}}{\mathrm{B}_0}\right)\cdot\nabla\delta p+\frac{1}{\mathrm{eB}_0}(\nabla\times\widehat{\mathbf{e}}_{||})\cdot\nabla\delta p+\frac{2\mu_0}{\mathrm{eB}_0{}^3}(\widehat{\mathbf{e}}_{||}\times\nabla P_0)\cdot\nabla\delta p \tag{6.19}$$

where we started by assuming $\delta\mathrm{B}_{||}<<\mathrm{B}_0$, then used quasimeutrality, linearized and finally used (6.17) assuming that

$$\frac{\nabla\delta B_{||}}{\delta B_{||}}>>\frac{\nabla B}{B_0}$$

We shall now rewrite $\nabla\times\widehat{e}_{||}$ using standard vector relations

$$\nabla\cdot(\widehat{\mathbf{e}}_{||}\cdot\widehat{\mathbf{e}}_{||})=2\widehat{\mathbf{e}}_{||}\times(\nabla\times\widehat{\mathbf{e}}_{||})+2(\widehat{\mathbf{e}}_{||}\cdot\nabla)\widehat{\mathbf{e}}_{||}=0$$

Taking the vector product with $\widehat{\mathbf{e}}_{||}$ we find

$$(\nabla \times \widehat{\mathbf{e}}_{||})_{\perp} = \widehat{\mathbf{e}}_{||} \times (\widehat{\mathbf{e}}_{||} \cdot \nabla)\widehat{\mathbf{e}}_{||} \tag{6.20}$$

Since

$$\widehat{\mathbf{e}}_{||} \cdot (\nabla \times \widehat{\mathbf{e}}_{||}) = \frac{1}{B}\widehat{\mathbf{e}}_{||} \cdot (\nabla \times \mathbf{B})$$

is associated with a background current and since $k_{||}$ generally is assumed to be small we will neglect the parallel component of rot $\widehat{\mathbf{e}}_{||}$. We can now use (6.17) for background fields to express the first term in (6.19) in the two others. It then turns out that the finite beta terms cancel. Then introducing the curvature vector

$$\kappa = (\widehat{\mathbf{e}}_{||} \cdot \nabla\widehat{\mathbf{e}}_{||}) = -\frac{\mathbf{R_c}}{R_c^{\,2}}$$

We can write

$$\nabla \cdot \left[(n_i \mathbf{v}_{*i} - n_e \mathbf{v}_{*e}) \left(1 - \frac{\delta B}{B_0} \right) \right] \approx \frac{2}{eB_0} (\widehat{\mathbf{e}}_{||} \times \kappa) \cdot \nabla \delta p \tag{6.21}$$

This result has several interesting implications. First as we already noticed, the finite β terms cancel, making a low β treatment adequate. Second, we see that the divergence of the diamagnetic drift flux is a curvature effect (compare Eq. 2.12). The term given by (6.21) is in fact the leading order curvature effect in an expansion in a/R (inverse aspect ratio) and is the main driving pressure term for ballooning modes. It can be represented by an equivalent gravity drift and this gives the same result as obtained for the kinetic derivation of interchange modes. The drift terms $k_y v_g$ in the derivation are, however, higher order in a/R and do not correctly describe the effect of a curved field. The result (6.21) suggests that we introduce an effective curvature drift

$$\mathbf{v}_{\kappa j} \approx \frac{2T_j}{q_j B_0} (\widehat{\mathbf{e}}_{||} \times \kappa) \tag{6.22}$$

Which is the total magnetic drift in a Maxwellian plasma, including the lowest order finite β effects. When effects of $\delta B_{||}$ are not included we have the curvature relations

$$\nabla \cdot (n\mathbf{v}_*) = \frac{1}{T}\mathbf{v_D} \cdot \nabla \delta p \tag{6.23}$$

and

$$\nabla \cdot \mathbf{v}_E = \frac{q}{T}\mathbf{v_D} \cdot \nabla \phi \tag{6.24}$$

Where

$$\mathbf{v_D} = \mathbf{v}_\kappa + \mathbf{v}_{\nabla B} \tag{6.25}$$

$$\mathbf{v}_\kappa = \frac{\mathrm{T}}{\mathrm{m}\Omega_\mathrm{c}}(\widehat{\mathbf{e}}_{||} \times \kappa) \tag{6.26}$$

and

$$\mathbf{v}_{\nabla B} = \frac{\mathrm{T}}{\mathrm{m}\Omega_\mathrm{c}}\left(\widehat{\mathbf{e}}_{||} \times \frac{\nabla B}{B}\right) \tag{6.27}$$

is the sum of the grad B drift and the curvature drift. These drifts are the same as the kinetic grad B and the curvature drifts when those are averaged over a Maxwellian distribution, i.e., $<\mathrm{v}_{||}{}^2> = \mathrm{T}_{||}/\mathrm{m}$, $<\mathrm{v}_\perp{}^2> = 2\mathrm{T}_\perp/\mathrm{m}$, assuming isotropy. From the comparison we see that if the fluid equations were generalized to a situation with different $\mathrm{T}_{||}$ and $\mathrm{T}_\perp$ we should use $\mathrm{T}_\perp$ for the grad B drift and $\mathrm{T}_{||}$ for the curvature drift.

For unisotropic temperature we, in fact, get a contribution from the curvature drift to the fluid drift. It is [137]

$$\mathbf{v}_{Dfluid} = \frac{\mathrm{T}_{||} - T_\perp}{T_{||}}\mathbf{v}_\kappa \tag{6.28}$$

Where $\mathbf{v}_{\kappa=}\left(\mathrm{T}_{||}/\mathrm{m}\Omega_\mathrm{c}\right)\hat{e}_{||}\mathrm{x}\kappa$. Moreover the diamagnetic heat flow is split into two parts[137]:

$$\mathbf{q}_*{}^{||} = \frac{1}{2}\frac{P_\perp}{m\Omega_c}\widehat{\mathbf{e}}_{||} \times \nabla T_{||} + (P_{||} - P_\perp)\mathbf{v}_\kappa \tag{6.29}$$

$$\mathbf{q}_*{}^{\perp} = 2\frac{P_\perp}{m\Omega_c}\widehat{\mathbf{e}}_{||} \times \nabla T_\perp \tag{6.30}$$

For isotropic pressure these add up to the Braghinski **q.**

6.4 The Influence of Magnetic Shear on Drift Waves

As pointed out in the previous section, in a tokamak the magnetic field has both a toroidal and a poloidal component. Moreover, since the poloidal field is generated by the toroidal plasma current it varies with r. Assuming for instance a homogeneous current density and applying Ampéres law to a circular contour with radius r around the center of the plasma in the perpendicular plane we find $\mathrm{B_p} = (1/2)\mu_0\mathrm{jr}$

where j is the current density. It is thus natural to assume that B_p increases with r. In our previous Cartesian coordinate system the x coordinate corresponds to r and the y coordinate to the poloidal direction. The simplest possible approximation of the magnetic field is now:

$$\mathbf{B}(x) = B_0\left(\widehat{\mathbf{z}} + \frac{x}{L_s}\widehat{\mathbf{y}}\right) \tag{6.31}$$

where L_s is the characteristic scale length of the magnetic field variation. It usually fulfils $L_s/a >> 1$ where a is the small radius. This kind of transverse variation of the magnetic field is referred to as magnetic shear. In order to describe drift waves in a system with magnetic shear we have to solve a differential equation for the field variation in x, and the solution for the mode frequency becomes an eigenvalue problem. We now consider perturbations of the form:

$$f(x,y,z,t) = \widehat{f}(x)e^{i(k_y y + k_\parallel z - \omega t)} \tag{6.32}$$

Where f may represent any perturbed quantity. We may then write the perpendicular velocity, including $\mathbf{v}_E$ and v_{pi}, from (2.11c) and (2.11e) as:

$$\mathbf{v}_{\perp i} = \frac{1}{B_0}\frac{\partial\phi}{\partial x}\widehat{\mathbf{y}} - i\frac{k_y}{B_0}\phi\cdot\widehat{\mathbf{x}} + \frac{\omega}{B_0\Omega_{ci}}\left(i\frac{\partial\phi}{\partial x}\widehat{\mathbf{x}} - k_y\phi\cdot\widehat{\mathbf{y}}\right) \tag{6.33}$$

The ion continuity equation now yields

$$\frac{\delta n}{n_0} = \left[\frac{\omega_{*e}}{\omega} + \frac{T_e}{m_i\Omega_{ci}^{\ 2}}\frac{\partial^2}{\partial x^2} - \frac{k_y^{\ 2}T_e}{m_i\Omega_{ci}^{\ 2}} + \frac{T_e}{m_i}\frac{k_\parallel^{\ 2}}{\omega^2}\right]\frac{e\phi}{T_e} \tag{6.34}$$

We shall, for simplicity, disregard destabilizing effects and use the approximation:

$$\frac{\delta n_e}{n_0} = \frac{e\phi}{T_e} \tag{6.35}$$

for the electron density. We now want to introduce the leading order effect of the magnetic shear into the system (6.33) and (6.34). The effect of the magnetic shear will be to twist the magnetic field. A toroidal eigenmode will also be twisted according to its poloidal and toroidal mode numbers. At a certain value of r it has the same degree of twisting as the magnetic field and $k_\parallel = 0$. At larger r the poloidal field will have a projection on z. The simplest model for its variation in a Cartesian system is (compare Eq. 6.6).

$$k_\parallel = \frac{x}{L_s}k_y \tag{6.36}$$

Now combining (6.34) and (6.35), using the quasineutrality condition we obtain the eigenvalue equation

$$\rho_s{}^2 \frac{\partial^2 \phi}{\partial x^2} + \frac{c_s{}^2}{\mathrm{v}_{*e}{}^2} \frac{x^2}{L_s{}^2} \phi + \left[\frac{\omega_{*e}}{\omega} - 1 - k_y{}^2 \rho_s{}^2\right] \phi = 0 \tag{6.37}$$

where we approximated ω by ω_{*e} in the term proportional to x^2, since this term is assumed to be small. Equation 6.37 has a solution of the form

$$\phi = H_n(i\xi) e^{\pm i\xi^2/2} \tag{6.38}$$

Where H_n is a Hermite polynomial of order n and

$$\xi = \left(\frac{\Omega_{ci}}{\mathrm{v}_{*e} L_s}\right)^{1/2} x$$

If (6.38) is substituted into (6.37) we obtain the condition

$$\frac{\mathrm{v}_{*e} \Omega_{ci} L_s}{c_s{}^2} \left[\frac{\omega_{*e}}{\omega} - 1 - k_y{}^2 \rho_s{}^2\right] = \pm(2n + 1) \tag{6.39}$$

which determines the eigenvalue ω. Clearly the $\pm$ in (6.38) and (6.39) is related to the direction of propagation of the wave. Assuming absorbing boundaries the group velocity must be outward. Since this corresponds to an inward phase velocity we have to choose the minus sign in (6.38). This leads to a convective damping for waves with outgoing group velocity. The mode which is easiest to destabilize is $n = 0$. For this mode the solution is [12]:

$$\omega = \omega_{*e} \left(1 - k_y{}^2 \rho_s{}^2\right) \left(1 - i\frac{L_n}{L_s}\right) \tag{6.40}$$

This case corresponds to

$$\phi = \Gamma e^{-i\xi^2/2} \tag{6.41}$$

where Γ is a constant. As we found previously drift waves have the strongest tendency for instability for small $k_\parallel$ where the electron shielding is inefficient. We thus expect drift waves to be generated near $k_\parallel = 0$ and then propagate towards larger x. When $k_\parallel$ has grown so that $k_\parallel \mathrm{v}_{ti} = \omega$ the ion-Landau damping sets in and absorbs the wave, thus preventing reflection at the plasma boundary and justifying the outgoing boundary condition. The extent of the wave in the x direction, due to the limiting effect of ion-Landau damping can be estimated to be

$$\lambda_x = \frac{\mathrm{v}_{*e}}{\mathrm{v}_{ti}} L_s$$

In order to have an absolute instability the growthrate of a drift instability must exceed the damping due to convection. It has recently been shown both analytically and numerically that if fully kinetic models are used both the collisional and the universal drift instabilities are only convective. In toroidal systems with poloidal variation of B_0, however, toroidal couplings may introduce absolute instability [26].

6.5 Interchange Perturbations Analysed by the Energy Principle Method

As mentioned in the first section one common method of determining the stability of a system is by calculating the change in energy caused by a small perturbation. We will here apply this method to an exchange of flux tubes (a tube where no magnetic field lines are crossing the mantle surfaces). Since the most unstable perturbations are electrostatic (no bending of field lines, $k_{\|} = 0$) we will consider only electrostatic perturbations.

For a Maxwellian velocity distribution the average particle energy is

$$E = \frac{1}{2} NT$$

Where N is the number of degrees of freedom. The equation of state is written

$$p = C(nm)^{\gamma}$$

Where

$$\gamma = \frac{2+N}{N} \quad \text{or} \quad N = \frac{2}{\gamma - 1}$$

Accordingly

$$E = \frac{T}{\gamma - 1}$$

and the internal energy in a volume v is

$$W_p = n\text{v}\frac{T}{\gamma - 1} = \frac{p\text{v}}{\gamma - 1} \tag{6.42}$$

where n is the particle density and $p = nT$ is the pressure.

We will now consider the exchange of plasma and magnetic flux from volume 1 into volume 2 and vice versa according to Fig. 6.2. Assuming an adiabatic process

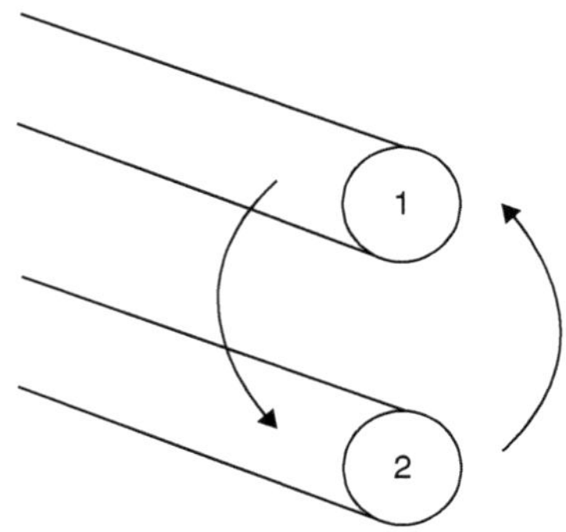

Fig. 6.2 Interchange of flux tubes

$$\frac{d}{dt}(p\mathrm{v}^{\gamma}) = 0 \tag{6.43}$$

The change in energy can be written

$$\Delta W_p = \frac{1}{\gamma - 1}\left[p_1\left(\frac{\mathrm{v}_1}{\mathrm{v}_2}\right)^{\gamma}\mathrm{v}_2 + p_2\left(\frac{\mathrm{v}_2}{\mathrm{v}_1}\right)^{\gamma}\mathrm{v}_1 - p_1\mathrm{v}_1 - p_2\mathrm{v}_2\right] \tag{6.44}$$

Where we used the relation (6.43) in the form

$$p_1{\mathrm{v}_1}^{\gamma} = p_2{\mathrm{v}_2}^{\gamma}$$

For small perturbations we may write $\mathrm{p}_1 = \mathrm{p}, \mathrm{v}_1 = \mathrm{v}\ \mathrm{p}_2 = \mathrm{p} + \delta\mathrm{p}$ and $\mathrm{v}_2 = \mathrm{v} + \delta\ \mathrm{v}$ where $\delta\mathrm{p} << \mathrm{p}$ and $\delta\mathrm{v} << \mathrm{v}$. Introducing these expressions into (6.44) we obtain

$$W_p = \delta p \delta \mathrm{v} + \gamma p \frac{\delta \mathrm{v}^2}{\mathrm{v}} \tag{6.45}$$

Since the second term is always positive a sufficient condition for stability is

$$\delta p \delta \mathrm{v} > 0 \tag{6.46}$$

We may write $\varphi = \mathrm{BS}$ where S is the surface of the cross-section of the flux tube. Since φ is constant along the flux tube we may write

$$\delta \mathrm{v} = \delta \int S dl = \phi \delta \int \frac{dl}{B}$$

When flux tube 2 is closer to the plasma boundary than tube 1, $\delta\mathrm{p} < 0$. Then condition (6.46) becomes $\delta\mathrm{v} < 0$ or

$$\delta \int \frac{dl}{B} < 0 \tag{6.47}$$

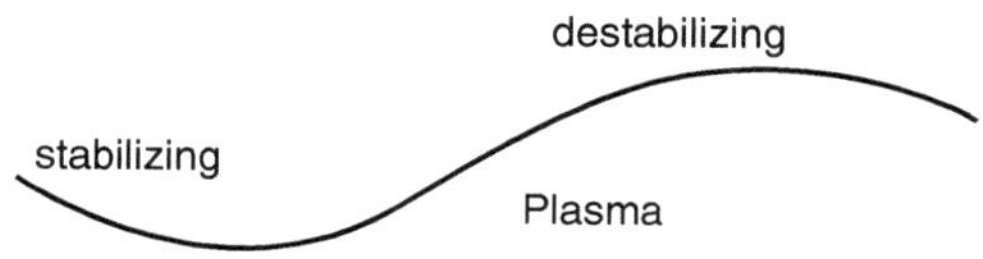

Fig. 6.3 Stabilizing and destabilizing curvature region

The condition (6.47) shows that configurations where the magnetic field on the average increases towards the plasma boundary are stable to flute perturbations $k_{\|} = 0$. Such configurations are denoted 'average minimum B systems'. For the simple case when B can be approximated by a vacuum field (low β) generated by an external current we have

$$B = \frac{2\mu_0 I}{R_c}$$

And the condition (6.47) takes the form

$$\delta \int R_c \, dl < 0 \tag{6.48}$$

showing that the plasma is stable when δR_c on the average is negative corresponding to a generating current situated in the direction of decreasing density. In this situation the magnetic field lines are concave into the plasma.

In the opposite case the contribution $\delta p \delta v$ to δ Wp is destabilising. In practice it turns out that such a system is normally unstable at least close to the boundary where p is small and accordingly also the second term in (6.45).

This interchange instability is equivalent to the instability described in Sect. 3.3.2 since the curvature of the magnetic field lines causes a centrifugal force that can be represented as an equivalent gravity

$$g = \frac{v_{th}^2}{R_c}$$

We then realise that when the curvature is destabilizing the gravity will be directed opposite to the density gradient corresponding to the necessary condition for instability κ g > 0 in (3.27b). (Fig. 6.3) Finally we emphasize once more that the condition (6.47) only says something about the average of the curvature along the field line. A real perturbation in a magnetic confinement device will experience a weighted average of the curvature which is determined by the mode structure and only if $k_{\|} = 0$ (flute mode) will the effective curvature be equal to the unweighted curvature giving the condition (6.47). Finite $k_{\|}$ modes will tend to become trapped in the destabilizing regions leading to a more unstable situation.

6.6 Eigenvalue Equations for MHD Type Modes

Since MHD type modes are more global than drift modes, a WKB approximation is often not valid and a careful inclusion of geometry is required. Thus we generally have to solve eigenvalue equations in the detailed geometry. We shall give examples here for simplified geometries, which nevertheless show the main features of the problem at the same time as an analytical description of the geometry is possible.

6.6.1 Stabilization of Interchange Modes by Magnetic Shear

As mentioned in Chap. 3, the electrostatic approximation for interchange modes has to be abandoned in a system with magnetic shear. We thus start from the description of electromagnetic interchange modes in Sect. 3.3.4 but now replacing the gravity drifts by the diamagnetic drifts. The condition div $\mathbf{j} = 0$ now takes the form

$$e\nabla \cdot [n_0 \mathbf{v}_{\mathrm{pi}} + n(\mathbf{v}_{*\mathbf{i}} - \mathbf{v}_{*\mathbf{e}})] = -\nabla \cdot (j_{||}\widehat{\mathbf{e}}_{||}) = \frac{1}{\mu_0}\nabla \cdot (\Delta_{\perp} A_{||}\widehat{\mathbf{e}}_{||}) \tag{6.49}$$

Here we make use of the approximation $\mathrm{E}_{||} = 0$ leading to

$$A_{||} = -\frac{i}{\omega}\widehat{\mathbf{e}}_{||} \cdot \nabla\phi \tag{6.50}$$

We shall, in the following, use a cylindrical coordinate system as in Sect. 6.2. The magnetic field will be written as

$$\mathbf{B} = B_\theta \widehat{\mathbf{e}}_\theta + B_\phi \widehat{\mathbf{e}}_\phi \tag{6.51}$$

where $\mathbf{e}_\theta$ and $\mathbf{e}_\varphi$ are unit vectors. Using the representation (6.14) for perturbations we find

$$\widehat{\mathbf{e}}_{||} \cdot \nabla f = \left[in\left(\frac{\nu}{r}B_\theta - \frac{1}{R}B_\phi\right) + \frac{B_\theta}{r\widehat{f}}\frac{\partial \widehat{f}}{\partial \theta}\right] f \tag{6.52}$$

Here ν is essentially the rotational transform q on a rational surface

$$\nu \approx \frac{B_\phi}{B_\theta}\frac{r}{R}$$

and we obtain

$$\widehat{\mathbf{e}}_{||} \cdot \nabla f \approx \frac{1}{\nu R}\left(\frac{1}{\widehat{f}}\frac{\partial \widehat{f}}{\partial \theta}\right)$$

Assuming R >> r we furthermore have

$$\Delta_{\perp}\widehat{f} \approx \left[-n^2\left(\int_0^{\theta}\frac{\partial \nu}{\partial r}d\theta\right)^2 - \frac{1}{r^2}n^2\nu^2\right]\widehat{f} \tag{6.53}$$

Where the r dependence of f was neglected, assuming large mode number n. If we neglect the θ dependence of ν (ν = q), (6.53) reduces to

$$\Delta_{\perp}\widehat{f} = -n^2\left[\left(\frac{dq}{dr}\right)\theta^2 + \frac{q^2}{r^2}\right]\widehat{f} = -n^2\frac{q^2}{r^2}(1 + s^2\theta^2)\widehat{f} \tag{6.54}$$

where

$$s = \frac{r}{R}\frac{dq}{dr}$$

The operator expression on the right hand side of (6.49) then takes the form

$$\nabla \cdot (\Delta_{\perp}A_{||}\widehat{\mathbf{e}}_{||}) = -\frac{i}{\omega}n^2\frac{q^2}{r^2}\frac{1}{q^2R^2}\frac{\partial}{\partial \theta}(1 + s^2\theta^2)\frac{\partial \phi}{\partial \theta} \tag{6.55}$$

For the divergence of the diamagnetic flux we use (6.18). This means that we take into account the lowest order finite β effect from $\delta B_{||}$ which enters only in this term. We then also have to know δp which to the lowest order can be taken as a convective perturbation, i.e.

$$\delta p = \frac{i}{\omega B_0}(\widehat{\mathbf{e}}_{||} \times \nabla \phi) \cdot \nabla P_0 = -\frac{i}{\omega B_0}(\widehat{\mathbf{e}}_{||} \times \nabla P_0) \cdot \nabla \phi \tag{6.56}$$

Expanding in r/R we now find

$$(\widehat{\mathbf{e}}_{||} \times \kappa) = \frac{1}{{B_0}^2}\left[\widehat{\mathbf{e}}_{||} \times (\mathbf{B} \cdot \nabla)\mathbf{B}\right] \approx \frac{1}{R}(\cos\theta\widehat{\mathbf{e}}_{\theta} + \sin\theta\widehat{\mathbf{r}} - \delta\widehat{\mathbf{e}}_{\theta})$$

where δ is an average part that is higher order in r/R. This expression gives the local curvature of the magnetic field lines entering (6.49). The part δ is the only remaining part when the integral is taken over the whole period in θ. We may express grad *f* as *i* **k***f* where

$$\mathbf{k} = n\left(\frac{dq}{dr}\theta r + \frac{q}{r}\widehat{\mathbf{e}}_\theta - \frac{1}{R}\widehat{\mathbf{e}}_\phi\right)$$

with the result

$$(\widehat{\mathbf{e}}_{||} \times \kappa) \cdot \mathbf{k} = \frac{nq}{rR}(\cos\theta + s\theta\sin\theta - \delta) \tag{6.57}$$

Now introducing (6.54), (6.55), (6.56) and (6.57) into (6.49) and replacing $\theta \rightarrow \eta$ where η is a generalized angle type variable we obtain the eigenvalue equation

$$\omega^2(1 + s^2\eta^2)\phi + \mathrm{k_c}^2\mathrm{v_A}^2\frac{\partial}{\partial\eta}(1 + s^2\eta^2)\frac{\partial\phi}{\partial\eta} + Dg(\eta)\phi = 0 \tag{6.58}$$

Where

$$\mathrm{k_c} = \frac{1}{qR} \qquad \mathrm{D} = 2\frac{T_e. + T_i}{m_i R}\frac{1}{P_0}\frac{dP_0}{dr} \tag{6.59}$$

and

$$\mathrm{g}(\eta) = \cos\eta + s\eta\sin\eta + \delta \tag{6.60}$$

The reason for introducing η is the ballooning mode formalism where it was found that η can take into account also the radial variation for large modenumbers.

Equation (6.58) represents the eigenvalue equation for electromagnetic interchange modes in a toroidal system with circular flux surfaces (**B** is assumed not to have an r component). The average curvature, δ is of order r/R and is not given with sufficient accuracy by the above treatment. We will here just regard it as a constant of order r/R, using expressions derived in the literature for various systems. As explained in the section on toroidal mode structure the relevant boundary condition for (6.58) in toroidal geometry is $\varphi \rightarrow 0$ as $\eta \rightarrow \infty$ thus including also part of the radial eigenvalue problem for large modenumbers. The interchange mode, which we will consider first as a highly elongated mode, will experience only the average curvature δ in (6.60). A common transformation for simplifying (6.58) is

$$\Psi = (1 + s^2\eta^2)^{\frac{1}{2}}\phi \tag{6.61}$$

Leading to the eigenvalue equation

$$\frac{\partial^2\Psi}{\partial\eta^2} + \left[\Omega^2 - \frac{s^2}{(1 + s^2\eta^2)^2} - \frac{\alpha\delta}{1 + s^2\eta^2}\right]\Psi = 0 \tag{6.62}$$

Where $\Omega = \omega/(\mathrm{k_c v_A})$ and $\alpha = \mathrm{D}/(\mathrm{k_c v_A})^2$.

Approximate solutions of (6.62) can be obtained by substituting trial functions into the quadratic form

$$\int_0^{\infty} \left\{ \Psi \frac{\partial^2 \Psi}{\partial \eta^2} + \left[\Omega^2 - \frac{s^2}{(1+s^2\eta^2)^2} - \frac{\alpha\delta}{1+s^2\eta^2} \right] \Psi^2 \right\} = 0 \qquad (6.63)$$

The asymptotic solution to (6.62) for $\Omega^2 = 0$ is

$$\Psi = (1 + s^2\eta^2)^{1/2} (A\eta^{\gamma_1} + B\eta^{\gamma_2})$$

Where

$$\gamma_{1,2} = -\frac{1}{2}\left[1 \pm (1 + 4\alpha\delta/s^2)^{1/2}\right]$$

It can be shown that the probability of smoothly connecting this solution to the region $\eta \approx 0$, and at the same time making $\Omega^2 < 0$ in (6.56) depends on the sign of $1 + 4\alpha\delta/s^2$ giving stability when this expression is positive. The stability condition is thus

$$\frac{1}{2}s^2 + \alpha\delta > 0 \qquad (6.64)$$

Since $\alpha = -q^2 R d\beta/dr$ we now obtain the Mercier condition (3.32) for $\delta = (r/R)(1 - 1/q^2)$ and the Suydam criterion (3.31) for $\delta = -(r/R)q^2$ corresponding to toroidal and cylindrical geometry respectively.

6.6.2 *Ballooning Modes*

Another type of solution to (6.58) is a mode that varies strongly on the $\cos\eta$ space scale. Such a mode may localize in regions where the normal curvature $\cos\eta > 0$, thus experiencing unfavourable curvature on the average. For $s \sim 1$ it turns out that the $s\eta \sin\eta$ part of the curvature (named geodesic curvature) substantially extends the unfavourable curvature region. This is a ballooning mode. Since $\delta \sim a/R$ we will here neglect the average curvature. As it turns out, ballooning modes are also very sensitive to an η dependence of ν which is the lowest order effect in β of a deviation from circular flux surfaces. If we assume a harmonic variation of ν with η, i.e.

$$\nu(r, \eta) = q(r) + \hat{\nu}\cos\eta$$

we have to modify (6.54) to

$$\Delta_{\perp}\widehat{f}= -n^2\frac{q^2}{r^2}\left[1+\left(s\eta-\frac{\widehat{v}}{q}\sin\eta\right)^2\right]\widehat{f}$$

and (6.55) accordingly

We also find that (6.57) is replaced by

$$(\widehat{\mathbf{e}}_{||}\times\kappa)\cdot\mathbf{k}=\frac{nq}{rR}\left[\cos\eta+\left(s\eta\sin\theta-\frac{\widehat{v}}{q}\sin\eta\right)\sin\eta\right] \tag{6.65}$$

As can be shown by analytical solutions for the equilibrium at small β we have the relation $\widehat{v}/q=\alpha$. The eigenvalue equation for ballooning modes then takes the form

$$\Omega^2[1+(s\eta-\alpha\sin\eta)^2]\phi+\frac{\partial}{\partial\eta}[1+(s\eta-\alpha\sin\eta)^2]\frac{\partial\phi}{\partial\eta}+\alpha g(\eta)\phi=0 \tag{6.66}$$

Where

$$\mathrm{g}(\eta)=g^{(1)}(\eta)+g^{(2)}(\eta) \tag{6.67}$$

$$\mathrm{g}^{(1)}(\eta)=\cos\eta+s\eta\sin\eta \tag{6.68}$$

$$\mathrm{g}^{(2)}(\eta)=-\alpha\sin^2\eta \tag{6.69}$$

The eigenvalue equation (6.66) can be solved analytically for small s by deriving a quadratic form and using a trial function derived for small s and α by symmetric expansion. We again introduce the transformation (6.61). The lowest order eigenfunction can then be obtained by ignoring the slow sη dependence as:

$$\Psi^{(1)}=\frac{\alpha g^{(1)}}{1+s^2\eta^2}<\Psi> \tag{6.70}$$

where $<\Psi>$ has a slow background variation and Ω^2 is assumed to be small. The next order Ψ enters in the second harmonic equation

$$\Psi^{(2)}=\frac{\alpha^2}{4(1+s^2\eta^2)^2}[(1-2s^2\eta^2)\cos 2\eta+3s\eta\sin 2\eta]<\Psi> \tag{6.71}$$

The average Ψ, $<\Psi>$, asymptotically has to take the form $<\Psi>\sim e^{\pm i\Omega\eta}$, which is the same as would be obtained from (6.62). In the inner region we may for ballooning modes at small s make a constant approximation. The transition is estimated to occur at $\eta \sim 1/s$. we thus take the ansatz for $<\Psi>$ as:

$$<\Psi> = \begin{cases} 1 & \eta \leq \kappa/s \\ e^{i\Omega(\eta-\kappa/s)} & \eta \geq \kappa/s \end{cases} \tag{6.72}$$

where κ can, in principle, be determined by maximizing the growthrate variationally.

The result obtained from integrating a variational form in Ψ corresponding to (6.63) and using

$$\Psi = <\Psi> + \Psi^{(1)} + \Psi^{(2)}$$

can be written in the form

$$i(1+a)\Omega + \left(\frac{\kappa}{s} + b\right)\Omega^2 = \delta W \tag{6.73}$$

where δW is the energy change in dimensionless form given by

$$\delta W = \frac{\pi}{4s}\left[s^2 - \frac{3}{2}\alpha^2 s + \frac{9}{32}\alpha^4 - \frac{5}{2}\alpha e^{-1/s}\right] \tag{6.74}$$

Here the last term in (6.74) is due to a mixing of space scales $s\eta$ and $\cos\eta$ in the integration while the constants a and b are due to the overlapping of the space scales $s\eta$ and $i\Omega\eta$. Since a and b depend on κ in a rather complicated way a variational determination of κ is not practical. The solution for $< \Psi >$ can in principle be obtained from the 'averaged' equation

$$\frac{\partial^2 <\Psi>}{\partial\eta^2} + \left[\Omega^2 - \frac{s^2}{(1+s^2\eta^2)^2} + \frac{2\alpha s^2 - (3/8)\alpha^4}{(1+s^2\eta^2)^3}\right]<\Psi> = 0 \tag{6.75}$$

As can be verified numerically, the asymptotic solution $<\Psi> \sim e^{\pm i\Omega(\eta-\kappa/s)}$ holds essentially from the point of inflection for $<\Psi>$. This is in the centre of the unstable region given by $\alpha^2 \approx 8/3$ and $\Omega^2 \approx -\pi^2/16s^2$, $\eta_{infl} \approx 0.5/s$ where η_{infl} is the inflection point. Now, continuing the asymptotic solution to smaller η we realize that it will reach 1 somewhere in the interval $0 \leq \eta \leq \eta_{infl}$. The simplest possible choice is then to take the matching point in the middle of this interval, i.e.

$$\frac{\kappa}{s} = \frac{1}{2}\eta_{\inf l} = \frac{1}{4s} \tag{6.76}$$

Or $\kappa = 0.25$. This value gives good agreement between numerical and analytical results for small s. For $\kappa = 0.25$ we obtain:

$$a = \frac{\pi}{4}[-0.69 + 0.57\alpha^2/s - 0.11\alpha^4/s^2] \tag{6.77}$$

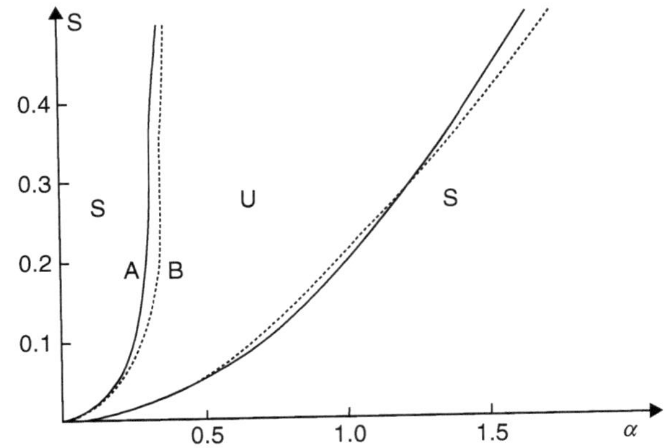

Fig. 6.4 Stability boundaries of the MHD ballooning mode. *A* numerical, *B* analytical (From [84], with the permission of the American Institute of Physics)

$$b = \frac{\pi}{4s}\left[0.3 - 0.18\alpha^2/s + 0.03\alpha^4/s^2\right] \tag{6.78}$$

As it turns out the solution of (6.73) is rather insensitive to the constants a and b, as it seems due to cancellation effects. The iΩ term in (6.73) is due to convective damping and represents the most important part of the frequency dependence in (6.73). When all terms are included extremely good agreement is obtained between the growthrate obtained from (6.73) and numerical results for small s. The agreement is, however, still within 20% in the centre of the unstable region for s ~ 0.25. The stability boundary as given by $\delta W = 0$ is shown in Fig. 6.4. We note the presence of the two stability regions, one for small α and one for large α. The stability for large α is due to the η dependence of ν. It is due to a reduction of the geodesic curvature due to finite pressure modification of the equilibrium. When the length of the destabilizing region decreases, the electromagnetic restoring force, through $k_{\|}$, has to increase.

If we include the lowest order FLR effect in a way corresponding to (3.44) we can simply make the substitution $\Omega^2 \rightarrow \Omega(\Omega - \Omega_{\bullet i})$. In this case the convective damping also influences the stability condition which takes the form:

$$\frac{1}{2}|\Omega_{*i}|(1+a) + \frac{1}{4}\left(\frac{\kappa}{s} + b\right)\Omega_{*i}{}^2 + \delta W > 0 \tag{6.79}$$

This condition should be compared to the condition (3.46) in the shearless case.

6.6.2.1 Kink Modes

While the interchange mode can be unstable inside the plasma (internal mode) the kink mode is more or less associated with the plasma boundary. It is due to a plasma current with a transverse gradient and can in the slab description easily be included as shown is Sect. 3.3.5 where it for a current profile extending over the whole cross

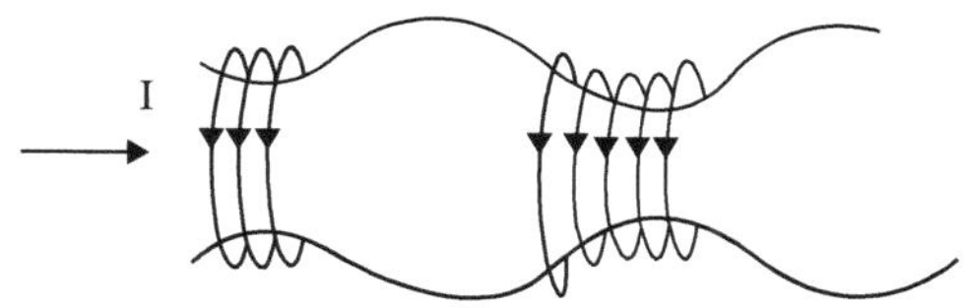

Fig. 6.5 The sausage instability corresponding to an m = 0 mode in a torus

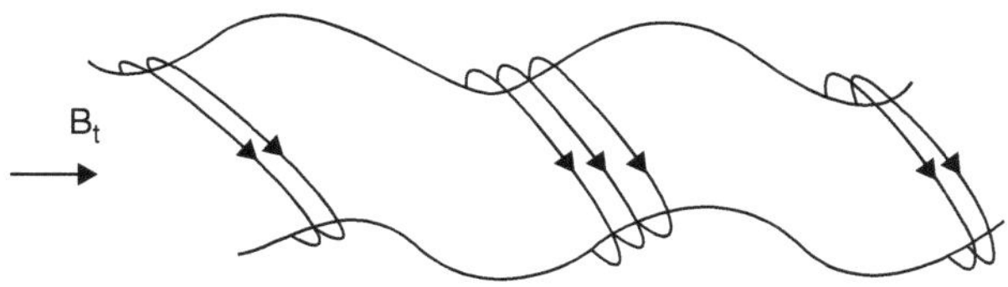

Fig. 6.6 A kink perturbation with m/n = Bt/Bp in a torus

section leads to a driving term $\Omega_{ci}k_{\|}v_0/n$ where v_0 is due to the background current and n is the toroidal modenumber. As a starting point we will consider the simple pinch in Fig. 6.5.

When the crossection of the current is decreased the magnetic field pressure increases and enhances the perturbation. For a toroidal configuration the pinch instability corresponds to m = 0 (no poloidal variation). In a system with toroidal magnetic field (along I in Fig. 6.5) the simple pinch instability is counteracted by the bending of the toroidal magnetic field lines. In a new configuration the total magnetic field will wind around the plasma in a way shown in Fig. 6.6. This system is now instead unstable to the perturbation shown in Fig. 6.6. Here we can see that again the instability occurs in such a way that a bending of the magnetic field lines is avoided. The new perturbation, however, has a finite poloidal variation determined by the relative magnitude of the poloidal, B_p and the toroidal, B_t, magnetic fields. This variation can be expressed by

$$k_p = k_t \frac{B_t}{B_p} \tag{6.80}$$

Modes with a slower poloidal variation are stabilized by the toroidal magnetic field while modes with a more rapid poloidal variation corresponding to a bending of the plasma current in the case shown in Fig. 6.6 can still occur. For a toroidal machine with $k_p = m/r$ and $k_t = n/R$ where r is the small radius and R is the large radius, the condition (6.80) becomes (compare the section on toroidal mode structure)

$$\frac{m}{r} = \frac{n}{R} \frac{B_t}{B_p}$$

While the stability criterion $k_p < k_t B_t / B_p$ becomes

$$q > \frac{m}{n} \tag{6.81}$$

Since q $\sim B_p^{-1} \sim I^{-1}$ we realize that (6.81) implies a limitation of the current I. Since $n \geq 1$ for kink modes, a sufficient condition for stability against mode m can be written $q > m$. For $m = 1$ this condition reduces to the Kruskal-Shafranov limit. The mode $m = 1$ is the least localized mode and extends over most of the crossection. It has also the largest growth rate, which for parabolic current profile can become a considerable fraction of the Alfv'en frequency for ballooning modes v_A/qR For larger m the kink modes become more and more localized to the plasma boundary.

We shall now make a more quantitative analysis of the kink mode, using a cylindrical geometry. This means that we use the representation (6.4)

$$f(r,\theta,\phi) = \widehat{f}(r)e^{i(m\theta - n\phi)}$$

neglecting the background inhomogeneity of the system in the θ direction. In this case the operators take the form

$$\nabla = \widehat{\mathbf{r}}\frac{\partial}{\partial \mathbf{r}} + i\frac{m}{r}\widehat{\mathbf{e}}_\theta - i\frac{n}{R}\widehat{\mathbf{e}}_\phi \tag{6.82}$$

$$\widehat{\mathbf{e}}_{||} \cdot \nabla = ik_{||}(r) = i\mathrm{k_c}(m - nq)$$

We shall in the following for brevity use the symbol $k_{||}$ for $k_c(m-nq)$, keeping its dependence on r in mind. As mentioned in Chap. 3 we may neglect the density perturbation from the polarization drift in (3.34). This equation is written in a general operator form and all we have to do is to replace z by a space dependent $\mathbf{e}_{||}$ and use the operator expression (6.76). we shall, however, also replace the gravity drifts by a real curvature, i.e. diamagnetic drifts with space dependent B_0 and $\mathbf{e}_{||}$. This leads to

$$\nabla \cdot [n(\mathbf{v}_{*\mathrm{i}} - \mathbf{v}_{*\mathbf{e}})] = \frac{2}{\mathrm{eB_0}}(\widehat{\mathbf{e}}_{||} \times \kappa) \cdot \nabla \delta p$$

Where

$$\kappa = (\widehat{\mathbf{e}}_{||} \cdot \nabla)\widehat{\mathbf{e}}_{||} = -\frac{\mathbf{R_c}}{\mathrm{R_c}^2}$$

Then using a convective pressure perturbation

$$\delta p = -\xi_\perp \cdot \nabla P_0$$

Where

$$\xi_\perp = -\frac{i}{\omega B_0}(\widehat{\mathbf{e}}_{||} \times \nabla\phi)$$

Leads to the result

$$\nabla \cdot [n(\mathbf{v}_{*i} - \mathbf{v}_{*e})] = 2i\frac{m^2}{r^2}\frac{1}{\omega e B_0{}^2 R_c}\frac{dP_0}{dr}\phi \tag{6.83}$$

We then arrive at the eigenvalue equation

$$\omega^2\frac{1}{r}\frac{\partial}{\partial r}r\frac{\partial\phi}{\partial r} - \left(\frac{m^2}{r^2}+\frac{n^2}{R^2}\right)(\omega^2 - k_{||}{}^2 v_A{}^2)\phi = k_{||}v_A\frac{1}{r}\frac{\partial}{\partial r}r\frac{\partial}{\partial r}k_{||}v_A\phi - \frac{B_0}{n_0 m_i}\frac{dJ_{||}}{dr}k_{||}\frac{m}{r}\phi$$
$$- 2\left(\frac{m}{r}\right)^2\frac{1}{m_i n R_c}\frac{dP_0}{dr}\phi \tag{6.84}$$

The question of stability is most easily studied in the energy integral formulation. Thus multiplying (6.84) by rφ and integrating from r = 0 to a, performing partial integrations of the terms containing (∂/∂r) r(∂/r) we obtain the energy formulation

$$\omega^2\int_0^a\left(r\frac{\partial\phi}{\partial r}\right)^2 + \int_0^a(\omega^2 - k_{||}{}^2 v_A{}^2)\frac{m^2}{r^2}\phi^2 r dr - v_A{}^2\int_0^a\left(\frac{d}{dr}k_{||}\phi\right)^2 r dr$$
$$- \frac{B_0}{n_0 m_i}\int_0^\infty m\frac{dJ_{||}}{dr}k_{||}\phi^2 dr = \left(\phi a\frac{d\phi}{dr} - k_{\{|}v_A\phi r\frac{d}{dr}k_{\{|}v_A\phi\right)_{r=a} \tag{6.85}$$

Where we neglected terms of order r/R. Since now

$$\int_0^a\left(\frac{d}{dr}k_{||}\phi\right)^2 r dr^2 = \int_0^a\left[k_{||}{}^2\left(\frac{d\phi}{dr}\right)^2 + \left(\frac{dk_{||}}{dr}\right)^2\phi^2 - \frac{1}{2}\phi^2\frac{1}{r}\frac{d}{dr}\left(r\frac{dk_{||}{}^2}{dr}\right)\right]rdr$$
$$-\frac{1}{2}\left(a\frac{dk_{||}{}^2}{dr}\phi^2\right)_{r=a}$$

We may write (6.85) for φ(a) = 0 (internal mode) as

$$\int_0^a\left[f(r)\left(\frac{\partial\phi}{\partial r}\right)^2 + g(r)\phi^2\right]dr = 0 \tag{6.86}$$

Where

$$f(r) = (k_{||}{}^2 v_A{}^2 - \omega^2)r \tag{6.87}$$

and

$$g(r) = \left(k_{\|}{}^{2}\mathrm{v_A}{}^{2} - \omega^{2}\right)\frac{m^{2}}{r^{2}} + \mathrm{v_A}{}^{2}\left(\frac{dk_{|}}{dr}\right)^{2} r - \frac{1}{2}\mathrm{v_A}{}^{2}\frac{d}{dr}\left(r\frac{dk_{\|}{}^{2}}{dr}\right) + \frac{B_0}{n_0 m_i}\frac{dJ_{\|}}{dr} m k_{\|} \tag{6.88}$$

If the terms proportional to ω^2 are separated out, the remaining terms are proportional to the energy change δW ($\omega^2<0$ corresponds to the unstable case where $\delta W<0$). The expression (6.88) for g(r) can be simplified if we make use of the relation between magnetic shear $dk_{\|}/dr$ and current. To lowest order in inverse aspect ratio the φ component of Ampéres law may be written

$$\frac{1}{2r}\frac{d}{dr}(rB_\theta) = \mu_0 J_\phi \approx \mu_0 J_{\|} \tag{6.89}$$

while

$$\frac{dk_{\|}}{dr} = m\frac{dk_c}{dr} = -mk_c\left(\frac{1}{r} - \frac{1}{B_\theta}\frac{dB_\theta}{dr}\right) \tag{6.90}$$

Combining (6.89) and (6.90) we obtain

$$J_{\|} = \frac{1}{\mu_0}B_\theta\left(qR\frac{dk_{\|}}{dr} + \frac{2}{r}\right)$$

And accordingly

$$\frac{mB_0}{n_0 m_i}\frac{d}{dr}J_{\|} = \mathrm{v_A}{}^{2}\left(r\frac{d^2k_{\|}}{dr^2} + 3\frac{dk_{\|}}{dr}\right) \tag{6.91}$$

We then obtain

$$g(r) = \left(\omega^2 - k_{\|}{}^{2}\mathrm{v_A}{}^{2}\right)\frac{m^2}{r} - \mathrm{v_A}{}^{2}\left(\frac{dk_{\|}{}^{2}}{dr}\right) \tag{6.92}$$

Another simplification is obtained if we change the dynamic variable to $\xi = \phi/\mathrm{r}$ (the radial component of the plasma displacement is $\xi = 1/(\omega \mathrm{B})(\mathrm{m}/\mathrm{r})\phi$). We then obtain the energy formulation:

$$\int_0^a \left[f(r)\left(\frac{\partial\xi}{\partial r}\right)^2 + h(r)\xi^2\right] dr = 0 \tag{6.93}$$

Where

$$f(r) = r^3\left(k_{\|}{}^2 \mathrm{v_A}{}^2 - \omega^2\right)r$$

and

$$h(r) = (m^2 - 1)\left(k_{\|}{}^2 \mathrm{v_A}{}^2 - \omega^2\right)r$$

for stability we require $\omega^2>0$, and obtain the condition

$$\int_0^a (m - nq)^2 \left[\left(r\frac{\partial \xi}{\partial r}\right)^2 + (m^2 - 1)\xi^2\right] rdr>0 \tag{6.94}$$

This condition is fulfilled for modes with m > 1. If m = 1 a marginally stable mode with $d\xi/dr = 0$ can be constructed if q(0)<1 In this case higher order terms in r/R have to be included in order to determine stability.

For external modes, $\phi(a) \neq 0$, and appropriate boundary conditions have to be imposed at the plasma boundary. These are the conditions of pressure balance across the surface

$$\mathbf{B_0} \cdot (\delta \mathbf{B} + \xi \cdot \nabla \mathbf{B_0}) = cons\tan t$$

And the condition that the displaced plasma surface remains a flux surface

$$\delta \mathrm{B_r} = [\nabla \times (\xi_r \times \mathbf{B_0}]_r$$

Where $\boldsymbol{\xi}_r$ is the radial displacement,

If there is no stabilization due to a conducting wall this leads to the stability condition

$$\int_0^a \left(\frac{1}{q} - \frac{m}{n}\right)^2 \left[\left(r\frac{\partial \xi}{\partial r}\right)^2 + (m^2 - 1)\xi^2\right] rdr$$
$$+ \left[\frac{2}{q_a}\left(\frac{n}{m} - \frac{1}{q_a}\right) + (1 + m)\left(\frac{n}{m} - \frac{1}{q_a}\right)^2\right] a^2 \xi_a>0 \tag{6.95}$$

Where index a indicates the value at r = a. The condition (6.95) can be violated only if nq_n < m, i.e. a condition equivalent to (6.81) evaluated at the plasma boundary.

Another way of writing the stability condition is by using the relation (6.89) in the other direction, i.e. expressing all effects of magnetic shear in $J_{\|}$. This leads to the condition

$$\int_0^a \left[\frac{1}{\mu_0}\delta B^2 + B_\theta\left(1 - \frac{nq}{m}\right)\frac{dJ_{||}}{dr}\xi_r^{\ 2}\right] rdr > 0 \tag{6.96}$$

which shows that for $dJ_{||}/dr<0$ which is the usual case, only the regions where $nq<m$ are destabilizing.

We finally also note that we can use (6.91) in order to rewrite (6.84) in the form

$$\begin{aligned}\frac{\partial}{\partial \mathrm{r}}\left[\left(k_{||}^{\ 2}\mathrm{v_A}^2 - \omega^2\right) r\frac{\partial\phi}{\partial r}\right] + \frac{m^2}{r^2}\left(\omega^2 - k_{||}^{\ 2}\mathrm{v_A}^2\right)\phi - \mathrm{v_A}^2\frac{d^2 k_{||}^{\ 2}}{dr^2}\phi \\ + \frac{m^2}{r}\frac{B_0}{2n_0 m_i R}\frac{dP_0}{dr}\phi = 0\end{aligned} \tag{6.97}$$

Equation 6.97 agrees with (19a) in Ref. [38] if the displacement $\xi \sim \phi/\mathrm{r}$ is introduced. An important property of (6.97) is the presence of singularities when $k_{||}^{\ 2}\mathrm{v_A}^2 = \omega^2$. Assuming that near such a singularity $\partial\phi/\partial \mathrm{r} >> \phi/\mathrm{r}$ we may neglect all terms except the first in (6.97). This term can then be integrated to

$$\frac{\partial\phi}{\partial \mathrm{r}} = \frac{C}{r\left(k_{||}^{\ 2}\mathrm{v_A}^2 - \omega^2\right)}\phi \tag{6.98}$$

where C is a constant of integration, thus justifying our WKB approximation close to $k_{||}^{\ 2}\mathrm{v_A}^2 = \omega^2$ and showing the presence of a singularity. Since $k_{||}\mathrm{v_A}$ is usually a monotonous function of r we may have solutions in a continuous range of ω with the location of the singularity varying with ω. This continuous range of solutions is usually referred to as the *Alfvén continuum*. In the presence of toroidicity there will, however, exist a minimum in $k_{||}^{\ 2}\mathrm{v_A}^2$. Then in the region where ω^2 is smaller than this minimum there is no singularity and the eigenvalues of ω form a discrete spectrum. These modes are referred to as *global modes* since they are not restricted in space by a singularity.

6.7 Trapped Particle Instabilities

In a tokamak, the magnetic field consists of a toroidal (along the torus) and a poloidal (around the cross-section) component. Thus the magnetic field lines are wound in the way shown in Fig. 6.7

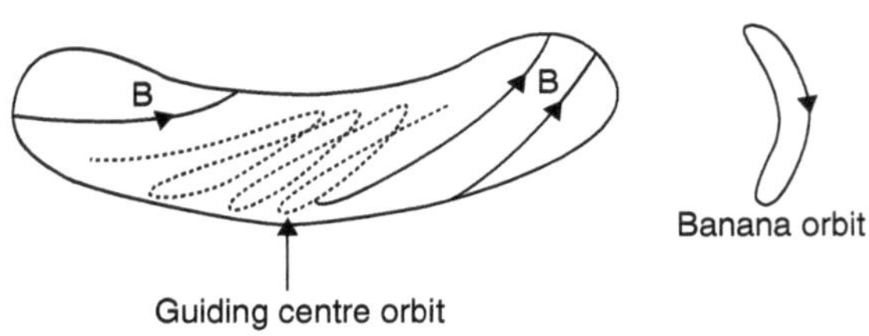

Fig. 6.7 Trapped particle orbits

When the small radius a is much smaller than the large radius R ($a << R$) the magnetic surface, defined as the surface described by the field line during many turns around the torus, has an almost circular cross-section. The toroidal magnetic field, however, is decreasing in the direction from the centre (along R). This means that a magnetic mirror is formed. A particle with small enough velocity along B will then be trapped in such a way that it never reaches the point $\theta = \pi$. In fact, it turns out that particles with $v_{\parallel} < (\sqrt{2})\varepsilon v_{\perp}$ where $\varepsilon = r/R$ become trapped. Assuming an isotropic velocity distribution function we then conclude that a fraction $(\sqrt{2})\varepsilon$ of the particles will be trapped. Clearly such trapping effects may decrease the possibilities for electrons to cancel space charge by moving along the magnetic field. Another important effect of the trapping is, however, to increase the effective collision frequency. Normally the collision frequency ν corresponds to 90° scattering. For trapped particles, however, clearly a scattering angle of $\sqrt{\varepsilon}$ is very significant since it may lead to detrapping. This may be accounted for by introducing a collision frequency $\nu_{eff} = \nu/\varepsilon$.

When studying systems with trapped particles we have to treat trapped and untrapped particles separately. For the study of trapped electrons we use the drift kinetic equation (5.7) in the electrostatic approximation. We will, however, include a magnetic mirror force as can be obtained by including (5.8) in (5.4). Assuming an unisotropic Maxwellian distribution with $T_{\perp} >> T_{\parallel}$ (relevant for trapped particles) we can see that the $\partial f/\partial v_{\perp}$ part can be neglected. We then obtain, including a Krook collision term

$$\begin{aligned}&\frac{\partial f}{\partial t} + ik_{\parallel}v_{\parallel}f - i\frac{1}{B_0}k_{\theta}\phi\frac{\partial f}{\partial r} + \left(\frac{e}{m}ik_{\parallel}\phi - \frac{\mu}{m}\frac{\partial B_0}{\partial z}\right)\frac{\partial f_0}{\partial v_{\parallel}}\\&= -\nu_{eff}\left(f - \frac{e\phi}{T_e}f_0\right) = 0\end{aligned} \tag{6.99}$$

The collision term relaxes the distribution function to a Maxwellian at potential φ in a time ν_{eff}^{-1}. In the force along B_0 we have included the effect of the inhomogeneity of B. This force is proportional to the magnetic moment μ. In order to have a strong influence of the trapping we realize that we must assume $\omega_B >> \omega$ where ω_B is the bounce frequency of particles due to trapping. If this condition is fulfilled the trapping may prevent the thermalization of the particles in the wave field. The particles then see a stationary field during a bounce period. Since then for a closed orbit a contribution $v_{\parallel}f_{Te}dt$ to the orbit integral will be cancelled by an equal contribution where $v_{\parallel} \rightarrow -v_{\parallel}$, we realize that the orbit average

$$\int_0^{2\pi/\omega_B} v_{\parallel}f_{Te}\,dt = 0$$

Introducing

$$\frac{\partial f_{0Te}}{\partial \mathrm{v}_{||}} = -\frac{m_e}{T_e}\mathrm{v}_{||} f_{0Te}$$

we find that the orbit average of the fourth term is also zero. This follows from the fact that the energy exchange $\int F v_{||} dt$ is zero where F is the force in (6.99). We then arrive at the averaged equation

$$\frac{\partial f_{Te}}{\partial t} - \frac{i}{B_0} k_\theta \phi \frac{\partial f_{Te}}{\partial r} = -\nu_{eff}\left(f_{Te} - \frac{e\phi}{T_e} f_{0Te}\right) = 0 \tag{6.100}$$

Since we have now removed all explicit $v_{||}$ dependence we can integrate (6.100) over $v_{||}$, thus replacing f_{Te} by $\delta\, n_e$ and f_{0Te} by n_{0Te} . Now introducing and

$$\frac{\partial f_{0Te}}{\partial \mathrm{r}} = \kappa f_{0Te}$$

We obtain

$$-\,i(\omega + i\nu_{eff}) f_{Te} + i(\omega_{*e} + i\nu_{eff}) \frac{e\phi}{T_e} f_{0Te} = 0$$

Then, considering the relation integrated over $v_{||}$, we have the trapped electron perturbation

$$\frac{\delta n_{Te}}{n_{0Te}} = \frac{\omega_{*e} + i\nu_{eff}}{\omega + i\nu_{eff}} \frac{e\phi}{T_e} \tag{6.101}$$

Assuming that free electrons thermalize (reach a Boltzmann distribution), we arrive at the electron density

$$\frac{\delta n_e}{n_0} = \left[\sqrt{\varepsilon}\frac{\omega_{*e} + i\nu_{eff}}{\omega + i\nu_{eff}} \frac{e\phi}{T_e} + 1 - \sqrt{\varepsilon}\right] \frac{e\phi}{T_e} \tag{6.102}$$

Where $\sqrt{\varepsilon}$ is the fraction of trapped electrons. If the bounce frequency of the ions, ω_{Bi} fulfills $\omega_{Bi} << \omega$ we may disregard the effect of trapping on ions. The ion density response is then:

$$\frac{\delta n_i}{n_0} = \left(\frac{\omega_{*e}}{\omega} - k_\theta{}^2 \rho_s^2 + \frac{k_{||}{}^2 c_s{}^2}{\omega^2}\right) \frac{e\phi}{T_e} \tag{6.103}$$

Now, using quasineutrality and treating $k_\theta{}^2\rho_s{}^2$, $k_{||}{}^2 c_s{}^2$ (but not $\sqrt{\varepsilon}$) as small we arrive at the dispersion relation

$$\omega \approx \omega_{*e}\left(1 - \frac{k_\theta{}^2\rho_s^2 - k_\parallel{}^2 c_s{}^2/\omega_{*e}{}^2}{1-\sqrt{\varepsilon}}\right) + i\frac{\nu_{eff}\sqrt{\varepsilon}}{\omega_{*e}(1-\sqrt{\varepsilon})}(\omega - \omega_{*e}) \tag{6.104}$$

Assuming a solution $\omega = \omega_r + i\gamma$, where $\gamma << \omega_r$ we now find the growthrate

$$\gamma = \frac{\nu_{eff}\sqrt{\varepsilon}}{(1-\sqrt{\varepsilon})^2}\left(k_\theta{}^2\rho_s^2 - k_\parallel{}^2 c_s{}^2/\omega_{*e}{}^2\right) \tag{6.105}$$

We thus find that the growthrate is modified by the factor $(\sqrt{\varepsilon})(1-\sqrt{\varepsilon})^{-2}$ in addition to the effect of trapping on the effective collision frequency, $\nu_{eff} = \nu/\varepsilon$. This instability is the trapped electron instability. When $\omega_{Bi} > \omega$ we may also have a trapped ion instability. Because the trapped particle distribution behaves as if $k_\parallel = 0$, i.e. as for a flute mode, the trapped particle modes may also be driven unstable by a magnetic curvature.

In the presence of an electron temperature gradient a new branch of this mode is introduced by trapping. This mode is believed to be responsible for the Alcator scaling of the energy confinement time in a tokamak.

6.8 Reactive Drift Modes

The eigenmodes that we have considered till now have basically either been of drift type, characterized by nearly Boltzmann distributed electrons, or of the MHD type characterised by small or zero parallel electric field. As shown in Chap. 3, the MHD modes are, in general, of a more global character and often show reactive instability i.e. instability without dissipative effects. The drift modes, on the other hand, in general require dissipation to become unstable. The reason for this is that Boltzmann distributed electrons are free to move along field lines to cancel space charge. Accordingly the charge separation caused by gravity or magnetic drifts is cancelled and the interchange instability does not occur.

There exists, however, also a third class of modes between the MHD modes and the usual drift waves. This class may be called reactive drift modes and typically has the maximum growthrate for $k^2\rho^2 \sim 0.1$. Since the ideal MHD modes generally have to be stable in fusion machines the reactive drift modes which are the second most dangerous class of modes are the potentially most likely candidates for explaining the observed transport in tokamaks. The first derivation of this new class of modes, was made by Rudakov and Sagdeev 1961 [1] when they discovered the slab η_i mode (6.120). This was, in fact, also the first work on drift waves as a whole. Later the trapped particle modes which also belong to this class were discovered by Kadomtsev and Pogutse [10].

6.8.1 Ion Temperature Gradient Modes

There are basically two ways through which a reactive instability can be recovered for drift waves. The first has already been indicated in the previous section. If a part of the electrons are trapped they will not be able to cancel space charge and an interchange instability is recovered. As a second possibility we notice that when a real curvature is used, an interchange mode is driven by the full pressure gradient [see (6.68)]. Here the temperature gradient part does not correspond to a charge separation but rather a compressibility. In the fluid sense a compressibility comes about as a divergence of a velocity. A velocity with a divergence has to vary in its own direction, thereby causing local rarefractions and bunchings. Since the convective part of div(n**v.**) is cancelled by a part of div $\mathbf{v}_\pi$, the full driving pressure term appears as a compressibility. If we, however, replace **v.** by a gravity drift where the temperature is perturbed, it becomes clear that it is the temperature perturbation part which is associated with compressibility

$$\nabla \cdot (n\mathbf{v}_g) = \mathbf{v}_g \cdot \nabla n + n\nabla \cdot \mathbf{v}_g$$

where now

$$\nabla \cdot \mathbf{v}_g = \frac{1}{T}\mathbf{v}_g \cdot \nabla \delta T$$

When δT is due to E × B convection in a background gradient, i.e.

$$\delta T = -\eta \frac{\omega_{*e}}{\omega} q\phi \tag{6.106}$$

Where

$$\eta = \frac{d \ln T}{d \ln n} = \frac{L_n}{L_T}, \qquad L_n = -\left(\frac{1}{n}\frac{dn}{dr}\right)^{-1}, \qquad L_T = -\left(\frac{1}{T}\frac{dT}{dr}\right)^{-1}$$

we obtain the dynamics shown in Fig. 6.8. Here, as usual, the x and y directions correspond to the r and θ directions in a torus. The variation of v_g along its direction gives rise to a density perturbation. We now assume Boltzmann electrons (3.3)

$$\frac{\delta n_e}{n_e} = \frac{e\phi}{T_e} \tag{6.107}$$

while the ions are subject to the compressibility. Using quasi-neutrality we then obtain a feedback mechanism as shown in Fig. 6.9.

Whether there is a positive or negative feedback depends on the relative directions of g and grad T. Not surprisingly it turns out that the feedback is positive (destabilizing) when g and grad T have opposite directions i.e. in unfavourable curvature regions.

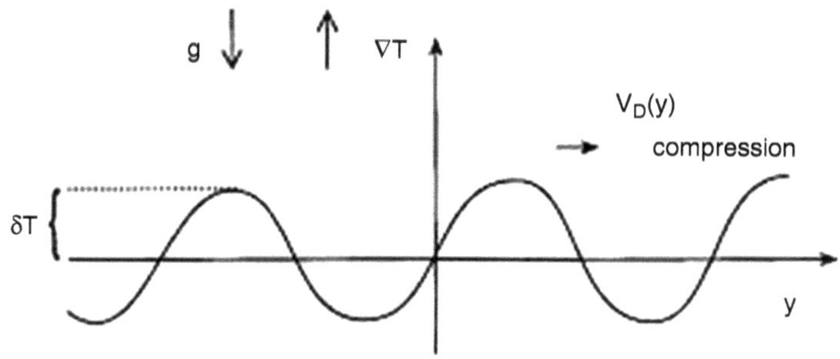

Fig. 6.8 Compression due to a magnetic drift

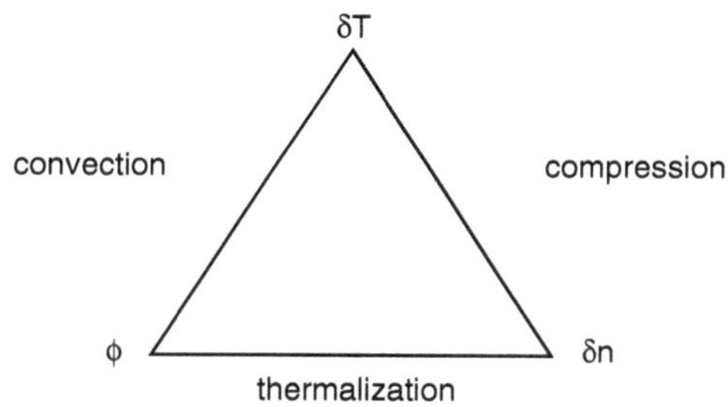

Fig. 6.9 Feedback loop of a thermal instability

Now combining (6.23) and (6.106) for the ions we obtain

$$\nabla \cdot (n\mathbf{v}_*) = \frac{1}{T}\mathbf{v}_{\mathbf{D}i} \cdot \nabla\left(T_i\delta n - \eta_i \frac{\omega_{*i}}{\omega} en\phi\right) \tag{6.108}$$

Also using (6.24) we now obtain from the ion continuity equation

$$\frac{\delta n_i}{n_0} = \frac{\omega_{*e} + \tau\omega_{Di} - \tau\eta_i(\omega_{Di}\omega_{*i}/\omega)}{\omega - \omega_{Di}} \frac{e\phi}{T_e} \tag{6.109}$$

When combined with (6.107) this gives the dispersion relation

$$\omega\left[\omega - \omega_{*e} - \omega_{Di}\left(1 + \frac{1}{\tau}\right)\right] = \eta_i\omega_{*e}\omega_{Di} \tag{6.110}$$

In the unstable case, (6.110) may be written $\omega = \omega_r + i\gamma$, where

$$\omega_r = \frac{1}{2}\left[\omega_{*e} - \omega_{De}\left(1 + \frac{1}{\tau}\right)\right] \tag{6.111}$$

$$\gamma = \omega_{*e}\varepsilon_n{}^{1/2}\sqrt{\eta_i - \eta_{ith}} \tag{6.112}$$

Where $\varepsilon_n = \omega_D/\omega_*$ and

$$\eta_{th} = \frac{1}{4}\left[1 - \varepsilon_n\left(1 + \frac{1}{\tau}\right)\right] \tag{6.113}$$

The present exercise merely serves to show that there is a reactive instability for large η_i. The magnetic drift terms have here not been treated consistently and several others should enter into (6.106) as will be shown later. The threshold (6.113) is thus incorrect. In early treatments the denominator in (6.109) was expanded for $\omega_D/\omega << 1$ and the ω_D/ω term combined with ω_{*e} in the numerator. This leads to

$$\frac{\delta n_i}{n_0} = \frac{1}{\omega}\left[\omega_{*e} - \tau(1 + \eta_i)(\omega_{Di}\omega_{*i}/\omega)\right]\frac{e\phi}{T_e} \tag{6.114}$$

where also the $\tau\omega_{Di}$ term is neglected as compared to ω_{*e}. This corresponds to using the convective density response directly in (6.108), i.e. the total pressure perturbation is convective. If now the stabilizing linear term in ω is ignored (6.114) leads to the stability threshold $\eta_i = -1$ which is often quoted in the literature. In this case part of the ω^2 term necessary for an instability has been obtained artificially by an expansion in ω_D/ω . This introduces a spurious instability for $\eta_i = 0$. The correct threshold is usually around $\eta_i = 1$ as will be shown later. The instability obtained here is a reactive drift instability driven by the temperature gradient and magnetic curvature. The mode is usually referred to as the toroidal η_i mode [28, 61, 74].

We have now shown how an instability is obtained when the compressibility originates from the divergence of $\mathbf{v}_{gi}$. The original η_i mode instability was, however, obtained due to the compressibility associated with the parallel ion motion. The feedback scheme in Fig. 6.9. applies also in this case.

For the parallel ion motion we take

$$\frac{\partial v_{\|i}}{\partial t} = -\frac{e}{m_i}\frac{\partial \phi}{\partial z} - \frac{1}{m_i n}\frac{\partial P_i}{\partial z} \tag{6.115}$$

Leading to

$$v_{\|i} = \frac{k_\|}{\omega m_i}\left(e\phi + \delta T_i + T_i\frac{\delta n}{n}\right) \tag{6.116}$$

Now using (6.106) for δT_i we obtain

$$v_{\|i} = \frac{k_\|}{\omega m_i}\left[\left(1 - \eta_i\frac{\omega_{*i}}{\omega}\right)e\phi + \delta T_i + T_i\frac{\delta n}{n}\right] \tag{6.117}$$

Including parallel ion motion in our derivation of the toroidal η_i mode we then obtain

$$\frac{\delta n_i}{n_0} = \left[\omega_{*e} + \tau\omega_{Di} - \tau\frac{\omega_{Di}\omega_{*i}}{\omega}\eta_i + \frac{k_{\|}{}^2 c_s{}^2}{\omega}\left(1 - \eta_i\frac{\omega_{*i}}{\omega}\right)\right]\left(\omega - \omega_{Di} - \frac{k_{\|}{}^2 c_s{}^2}{\tau\omega}\right)^{-1}\frac{e\phi}{T_e} \tag{6.118}$$

In order to consider the excitation due to parallel ion motion separately, we now put $\omega_D = 0$. The dispersion relation may then be written

$$\omega^3 - \omega^2\,\omega_{*e} - \omega k_{\|}{}^2 c_s{}^2\left(1 + \frac{1}{\tau}\right) + \eta_i\omega_{*i}k_{\|}{}^2 c_s{}^2 = 0 \tag{6.119}$$

The driving term here is the last term and the simplest possible dispersion relation giving the instability is

$$\omega^3 = -\eta_i\omega_{*i}k_{\|}{}^2 c_s{}^2 \tag{6.120}$$

Since $\omega_{*i}<0$, ω^3 is positive for positive η_i. Taking the phase angle as 2π we obtain an unstable root with phase angle $2\pi/3$. This instability, which does not require curvature, is usually referred to as the slab instability (slab mode) since its eigenvalue can be treated in slab geometry [1].

The η_i mode, is among the most serious candidates for explaining the anomalous ion heat transport in present day tokamaks. This may be anticipated already by its very fundamental nature as a thermal instability. When we heat a glass of water from below, we generate convection through a thermal instability. When we heat a tokamak with a centrifugal force due to field curvature we have a corresponding situation as when heating water and a similar thermal instability may develop. The toroidal version has the largest growth-rate in the bulk of the plasma while the slab version may have larger growth-rate close to the edge where $\omega_D << \omega_*$. The slab version usually has a slightly lower threshold while the parallel ion motion is stabilizing when the toroidal drive dominates [74, 92, 166]. Fully kinetic treatments show that both modes have their maximum growth-rate around $k_\perp{}^2\rho_i{}^2 = 0.1$.

6.8.2 Electron Temperature Gradient Mode

A mode that is sometimes used to try to explain the anomalous electron and heat transport in the collisionless regime is the electron temperature gradient mode (η_e mode). Also this mode exists in both slab (6.21) and toroidal [97, 115] versions. It is a very short wave-length mode fulfilling

$$\rho_e << \lambda << \rho_i$$

In this limit the ions are unmagnetized and furthermore in the hot regime. We may thus take

$$\frac{\delta n_i}{n} = -\frac{e\phi}{T_i} \tag{6.121}$$

The ions reach thermal equilibrium by moving perpendicular to the magnetic field. This also requires $\omega < k_\perp v_{thi}$ which can now also be fulfilled with $\omega > \Omega_{ci}$ The large mode number makes the frequency comparatively large. All that we will here require is that

$$\omega << \Omega_{ce} \tag{6.122}$$

In this regime we may still use the drift expansion for the electrons but it may be possible to ignore parallel electron motion. This means that as compared to the η_i mode the roles of ions and electrons are switched. We may thus follow the previous procedure. The electron density response (corresponding to (6.109)) is then

$$\frac{\delta n_e}{n_0} = \frac{\omega_{*e} - \omega_{De} + \eta_e(\omega_{De}\omega_{*e}/\omega)}{\omega - \omega_{De}} \frac{e\phi}{T_e} \tag{6.123}$$

In combination with (6.121) we now obtain the dispersion relation

$$\omega(\omega + \omega_{*e} - 2\omega_{De}) = -\tau\eta_e\,\omega_{*e}\,\omega_{De} \tag{6.124}$$

This dispersion relation is very similar to (6.110). An important difference is that the η_i mode propagates in the electron drift direction and the η_e mode propagates in the ion drift direction for small $|\omega_D|$. A correct treatment of the ω_D terms shows that for realistic values of ε_n the η_i mode propagates in the ion drift direction and the η_e mode in the electron drift direction. Such a trend can also be seen in our present treatment which is, however, not accurate enough to justify such a conclusion.

Due to the very short wave-length, the η_e mode only gives a small direct transport. It can, however, excite modes with longer wave-length through mode coupling. Such modes, with a wave-length of the order of the skin depth c/ω_{pe}. can give a neo Alcator scaling [cf Sect. 6.1]. The slab version of this mode is analogous to that of the η_i mode. We will not discuss it here.

6.8.3 Trapped Electron Modes

The most obvious candidates for explaining the large anomalous electron thermal conductivity in tokamaks are the trapped electron modes [10, 14]. As was mentioned previously, trapped electron modes can give a neo-Alcator scaling in the collision dominated regime. In the collisionless regime an interchange type of mode driven by the density gradient is often referred to as the Ubiquitous mode. We will here consider a collisionless trapped electron mode which is similar to the η_e mode but occurs for $k_\perp{}^2\rho_i{}^2 \sim 0.1$. Taking the fraction of the trapped electrons to be f_t we may use the response (6.123) for the trapped electrons since their motion along the magnetic field has been averaged out. We then have to consider the magnetic drift to be bounce averaged. The free electrons are assumed to be Boltzmann distributed. Then

$$\frac{\delta n_e}{n_0} = f_t \frac{\omega_{*e} - \omega_{De} + \eta_e(\omega_{De}\omega_{*e}/\omega)}{\omega - \omega_{De}} \frac{e\phi}{T_e} + (1 - f_t)\frac{e\phi}{T_e} \tag{6.125}$$

Now using (6.109), adding FLR effects for the ion response we obtain

$$\frac{\omega_{*e} + \tau\omega_{Di} - \tau\eta_i(\omega_{Di}\omega_{*e}/\omega) - k_\perp{}^2\rho_s{}^2(\omega - \omega_{*iT})}{\omega - \omega_{Di}} = \\ = f_t \frac{\omega_{*e} - \omega_{De} + \eta_e(\omega_{De}\omega_{*e}/\omega)}{\omega - \omega_{De}} + 1 - f_t \tag{6.126}$$

This relation shows a symmetry between ions and trapped electrons. We note that (6.126) is now a cubic equation in ω. This means that it has at least one real root and accordingly maximum two complex conjugate roots i.e. it can have no more than one unstable mode. The more complex fluid description in the next section gives a quartic equation and accordingly the possibility of having two unstable modes at the same time.

For that system it is possible to consider resonant modes where $\omega \sim \omega_D$ and in that way one may decouple the ion density perturbation. Here we will denote the left hand side of (6.126) by Δ. The dispersion relation for the trapped electron mode may then formally be written:

$$\omega\left(\omega + \frac{f_t}{\xi}\omega_{*e} - \frac{1-\Delta}{\xi}\omega_{De}\right) = -\frac{f_t}{\xi}\eta_e\omega_{*e}\omega_{De} \tag{6.127}$$

Where $\xi = 1 - f_t - \Delta$

Equation 6.127 is very similar to the dispersion relation (6.124) for the η_e mode. The dispersion relation shows only the electron dynamics and is a relevant description when the ion dynamics is subdominant. The ω^2 term is here entirely due to electron dynamics and we may have an instability driven only by electron compressibility and temperature gradient. This dispersion relation is accordingly analogous to (6.110) for the η_i mode which is destabilized by only ion dynamics. In fact, if we take the limit $\Delta \to 0$ in (6.127), the two modes are symmetric for $\tau = 1$ except for the factor f_t appearing in (6.127).

In order to investigate the other modes given by (6.126) we rewrite it in its cubic form

$$\omega(\omega-\omega_{De})(\omega-\omega_{*e}+\omega_{De}-\omega_{Di})-\omega\begin{bmatrix}\omega_{Di}\omega_{*e}\dfrac{\eta_i+\tau f_t\eta_e}{1-f_t}-\dfrac{k_\perp{}^2\rho_s{}^2}{1-f_t}(\omega-\omega_{De})\\(\omega-\omega_{*iT})\end{bmatrix}-$$
$$-\frac{f_t}{1-f_t}\omega(\omega_{Di}-\omega_{De})(\omega_{*e}-\omega_{De})=\frac{\eta_e f_t-\eta_i}{1-f_t}\omega_{*e}\omega_{De}\omega_{Di} \tag{6.128}$$

Here the single ω factor in the left hand side is associated with the temperature gradients and indeed if the temperature gradients vanish so does the right hand side, ω factors out and we obtain a quadratic dispersion relation. On the other hand we also note that the right hand side is quadratic in the magnetic drifts. We might thus neglect it for this reason thus obtaining a quadratic dispersion relation. Then neglecting also other terms that are quadratic in the magnetic drift we obtain the dispersion relation

$$(\omega-\omega_{De})\left[\omega\left(1+\frac{k_\perp{}^2\rho_s{}^2}{1-f_t}\right)-\omega_{*e}-\omega_{*iT}\frac{k_\perp{}^2\rho_s{}^2}{1-f_t}+\omega_{De}-\omega_{Di}\right]$$
$$=\frac{f_t}{1-f_t}\omega_{*e}(\omega_{Di}-\omega_{De})+\frac{\eta_i+\tau f_t\eta_e}{1-f_t}\omega_{*e}\omega_{Di} \tag{6.129}$$

We note that in the limit $f_t \to 0$ (6.129) is similar to (6.110) although the ω^2 term has a different origin. The differences are the inclusion of the FLR effect and the Doppler shift $\omega-\omega_{De}$ in the first ω factor. The latter difference is due to the nonadiabatic electron response which was absent in the derivation of (6.110). It is important to note that in the absence of both trapping and temperature gradients there is no instability i.e. the product of frequency independent parts in the left hand side of (6.162) cannot drive an instability. (Compare the discussion after (6.114)). Thus (6.129) is most conveniently solved by introducing $\acute{\omega}=\omega-\omega_{De}$ and first obtaining the solution for $\acute{\omega}$. In the absence of temperature gradients or compressibility (6.129) gives a pure trapped electron mode. This mode which may propagate either in the electron or ion drift direction depending on the values of $k_\perp{}^2\rho_i{}^2$ and ε_n is usually called the *ubiquitous mode* [19, 28]. The ω^2 term there requires the nonadiabatic responses from both ions and electrons. The ubiquitous mode is, in fact, stabilized by temperature gradients as we will see in Sect. 6.11. If we multiply (6.129) by $(1-f_t)$ and take the limit $f_t \to 1$ we obtain a pure MHD equation in the limit $k_\parallel = 0$.

6.9 Competition Between Inhomogeneities in Density and Temperature

As we have seen in (6.110) and (6.119) also very simple models for the temperature gradient mode indicate that the parameter $\eta = L_n/L_T$ is critical for stability. This is mainly because the diamagnetic drift ~$1/L_n$ introduces a real eigenfrequency that is stabilizing. A more fundamental reason for the importance of the parameter η is, however, the simultaneous convection in temperature and density gradients. This leads to a competition between convection and expansion (negative compression) in the energy equation. When the convection is outward higher density parts move out into more dilute areas where the expansion takes place. The expansion cools the plasma and competes with the increase in temperature due to temperature convection

$$\delta T = -\xi \cdot \nabla \mathrm{T} + \alpha \xi \cdot \nabla n$$

Here α is a coefficient that gives the expansion cooling. From this relation we immediately see that η has to exceed a certain limit for δT to be positive. A corresponding equation for n is obtained because a temperature perturbation, through $\mathbf{v}_D$ or $v_\parallel$ leads to compression. Thus

$$\delta n = -\xi \cdot \nabla \mathrm{n} + \beta \xi \cdot \nabla T$$

where we considered the convective temperature perturbation Fig 6.10. The competition between temperature and density gradients, in the nonlinear regime, leads to inward contributions to fluxes. When several sources of free energy are present (coupled relaxations) we may even have net inward fluxes (pinches) of some thermodynamic variables. The total energy flux is, however, always directed outward. A realistic threshold including the here discussed effects will be derived in the next section.

An important feature of feedback is that a negative feedback corresponds to a cooling in the direction of higher temperature. This would give a temperature pinch if the mode was not stable in this case. However, since the ExB convection is the same for ions and electrons the loops are coupled (Fig. 6.11) so a negative feedback in the ion loop may still give a pinch if the electron loop has positive feedback. This usually requires $\eta_e > \eta_i$. Of course the opposite may happen (electron temperature pinch if $\eta_i > \eta_e$). We can also replace the ion temperature loop with a density loop, shown in Fig. 6.12.

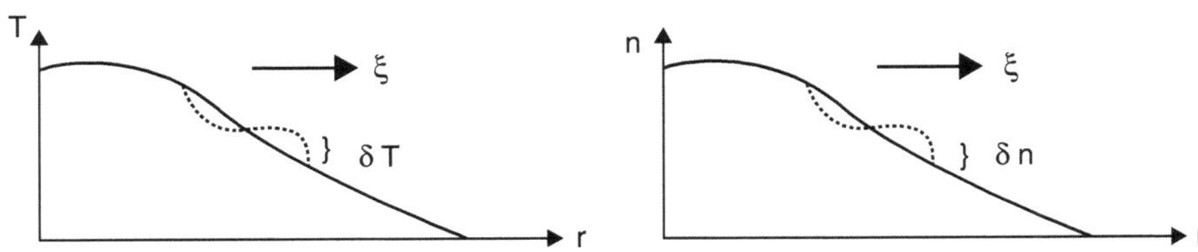

Fig. 6.10 Convective perturbations of temperature and density

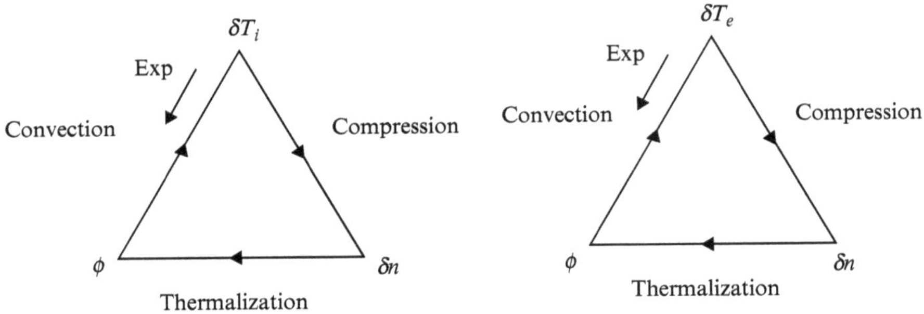

Fig. 6.11 Coupled loops of electron and ion temperature relaxation

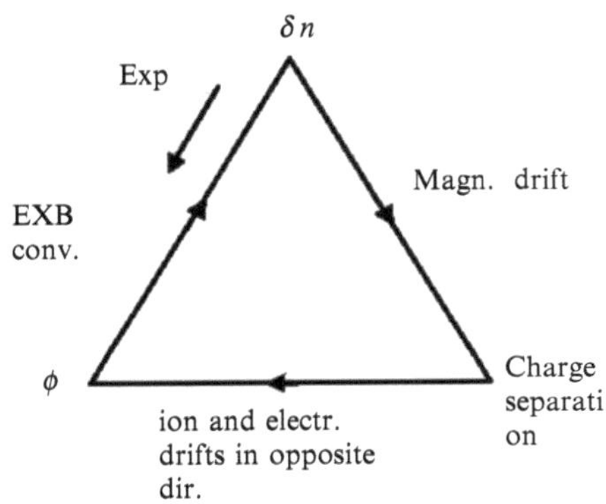

Fig. 6.12 Feedback loop of the ubiquitous mode

A density loop here corresponds to the feedback loop of the Ubiquitous mode which is a type of trapped electron mode. Thus the ubiquitous mode couples directly to the electron temperature dynamics and we can get a particle pinch if $\eta_e < 1$.

6.10 Advanced Fluid Models

One of the main problems with creating a first principles transport model, which can be used in transport codes, is the fact that because of the resonance

$$\omega = k_{||}v_{||} + \omega_D\left(v_{||}^2, v_{\perp}^2\right) \tag{6.130}$$

kinetic theory is, in principle, needed. On the other hand, not even the most efficient computers are able to run a fully nonlinear kinetic code as a part of a transport code. In fact, nonlinear kinetic simulations are usually made on time scales of the order linear growth time γ^{-1} and nonlinear saturation which is typically a few growth times while transport codes operate on time scales of the order of the confinement

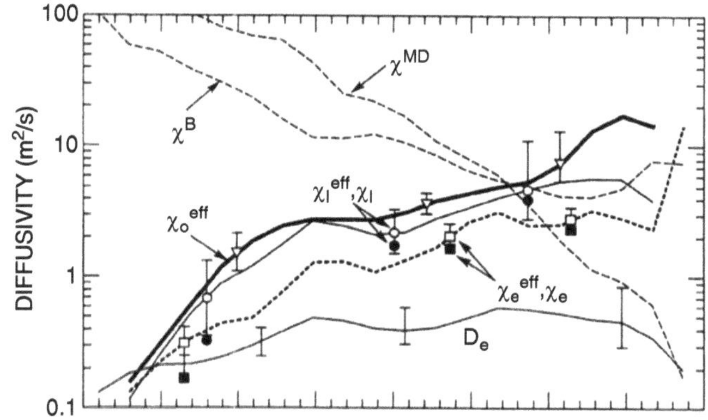

Fig. 6.13 Experimentally determined particle diffusivity, effective momentum and diffusivities for a TFTR discharge. The *solid symbols* on the curves for χ_i^{eff} and χ_e^{eff} show the corresponding values of χi and χe ($\chi_\varphi^{eff} = \chi_\varphi$ within ~10%). Also shown are the theoretical χφ (=χi) from Mattor and Diamond (Phys. Fluids 31, 1180 (1988)) (labeled χMD) and the χi from the toroidal ηi analysis of [103] (labeled χB) (Reprinted figure from [110], Fig. 2, with the permission of the American Physical Society (copyright 2008) and S.D. Scott). http://prola.aps.org/abstract/PRL/v64/i5/p531_1

time τ_E. While the growth time typically is of the order 10^{-5} s, τ_E is of the order of seconds. Thus, what we are left with for transport simulations is either kinetic models that ignore velocity space nonlinearities or some kind of advanced fluid models that attempt to incorporate the resonance (6.130) in some approximate way. The latter possibility has not been much explored until the end of the 1980s.

6.10.1 The Development of Research

The beginning of the development of advanced fluid theories, of course, depends on how we define the concept. With our definition, as will be given shortly, it dates back to 1986, with the first published papers appearing in 1987. Before this time all fluid models expanded the dynamic equations such that $\omega_D/\omega << 1$ (adiabatic state) for the perpendicular dynamics and introduced an equation of state with a free parameter γ that can describe adiabatic or isothermal states for the parallel motion. As it turns out, however, when kinetic or advanced fluid theories are used, ω and ω_D are usually comparable except at the edge. Because of this all previous drift-wave theories had a basic flaw in that transport coefficients decreased with radius in the models while they increased with radius in the experiments. As an example we may mention the work by Scott et.al. [110] where the radial profiles of ion thermal conductivity from 2 one-pole fluid models were compared with the experimental ion thermal conductivity for a TFTR shot (Fig. 6.13).

Although, in general, the parallel part of the resonance (6.130) may be important we will here first focus on the perpendicular part which is associated with the very fundamental toroidal effects. As a result of toroidicity, the eigenmodes tend to localize on the outside of the torus where the curvature is unfavourable. The ultimate limit, which when it can be reached gives the largest growthrate, is the local limit where the mode is strongly localized and effects of the parallel dynamics vanishes. In this limit the eigenvalue equation turns into an algebraic dispersion relation. Comparisons between local kinetic theory and a two-pole fluid model [108] showed that the diamagnetic heat flow $\mathbf{q}_{\boldsymbol{\cdot}}$ as given by (2.15), through the magnetic inhomogeneity part of its divergence, reproduces the main kinetic effects of ω_D in (6.130). By including this term in the fluid equations we obtain a two-pole fluid density response in the local limit and a three pole response in the 3 D case. It recovers both adiabatic and isothermal limits for both perpendicular and parallel dynamics. This was done in the advanced fluid model developed at Chalmers University of Technology in 1986 as described in Sect. 6.11.

It was first developed from the local limit of an electromagnetic model [90, 91] but later also the electrostatic eigenvalue problem was solved [92] using the Ballooning mode formalism, i.e. including also parallel ion dynamics id 3D. The ion thermal conductivity, based on quasilinear theory and mode-coupling simulations was published in 1988 [93, 94]. The increased order of the fluid response due to div $\mathbf{q}_{\boldsymbol{\cdot}}$ is significant since it changes the regions of positive and negative energy modes. This can be seen from the expression:

$$\varepsilon(\omega, \mathbf{k}) = \frac{1}{k^2 \lambda_{de}{}^2} \left(\frac{\delta n_i}{n} - \frac{\delta n_e}{n} \right) \left(\frac{e\phi}{T_e} \right)^{-1} \tag{6.131}$$

For the dielectric function in combination with the expression for the wave energy

$$W = \frac{\partial}{\partial \omega} (\omega \varepsilon(\omega, \mathbf{k})) |\nabla \phi|^2 \tag{6.132}$$

As an example we note that the electromagnetic version of this fluid model reproduces the instability of the MHD ballooning mode branch below the ideal MHD beta limit in the presence of an ion temperature gradient [70, 90]. This instability is due to div$\mathbf{q}_{\boldsymbol{\cdot}}$ and is caused by a shift of regions of negative and positive energy. From the drift wave point of view, div$\mathbf{q}_{\boldsymbol{\cdot}}$ introduces a new stability regime, with positive wave energy for large ε_n. This has the effect of giving a strong trend for χ_i to grow towards the edge as shown by Fig. 6.14. The new regime where ε_n is stabilizing is generally termed "the flat density regime". Since ε_n decreases towards the edge, the system departs more from marginal stability as we move towards the edge if the density and temperature profiles have similar shapes (Fig. 6.15).

The flat density regime typically prevails in the inner 80% of tokamak discharges which means that the new regime is dominant and radically changes the predictions of drift wave theory. The TFTR shot studied in [110] was also studied by the advanced fluid model described in [102], giving a χ_i which followed the experimental trend over the whole cross-section (Fig. 6.14). An upper stability regime in ε_n can also be obtained due to div $\mathbf{v}_E$ in an adiabatic model [98]. This stabilization is, however, of an FLR type, similar to that discussed in Sect. 4.5 since the wave

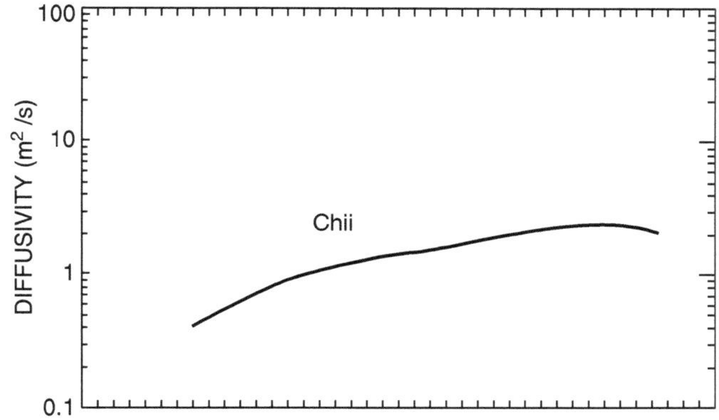

Fig. 6.14 Radial profile of χi using the advanced fluid model in [102] (From [193] with the permission of the AIP)

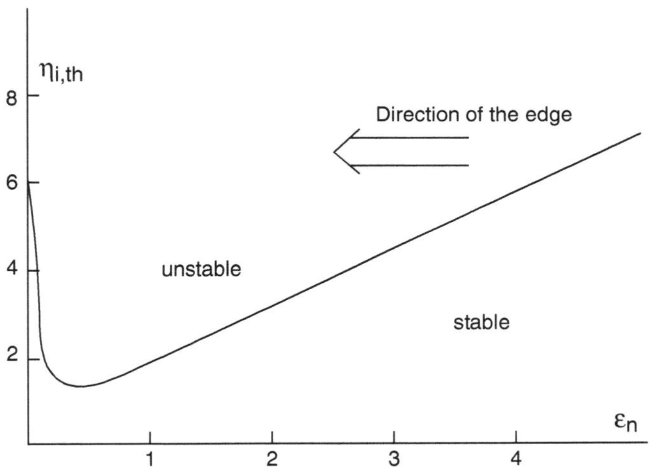

Fig. 6.15 Stability diagram in εn and ηi showing destabilization towards the edge

energy is still negative. As will be shown in Sect. 6.11.7. parallel ion motion destabilizes this regime by introducing a dissipative damping due to magnetic shear. We will thus introduce the following definition of an advanced fluid model:

> An advanced fluid model is a fluid model which recovers the stable regime of η_i modes for large ε_n due to positive wave energy.

In H-modes the density profile is flat over a large part of the discharge. This regime was investigated in [89] and it was pointed out that the critical parameter for stability here is R/L_{ti} rather than η_i. As it turns out most of the cross section of L-modes is also usually in the regime where this type of stability criterion applies. Such a regime was present in the general stability criterion of [91] but was not pointed out until in [93] and evaluated as $L_T/R = 0.367$. In [98] the FLR type stabilization in the adiabatic limit was obtained with the threshold $L_T/R = 0.28$ and in [108] the full local kinetic result $L_T/R = 0.35$ was obtained. It is interesting to note that a local kinetic model using a gradient B approximation of the magnetic drift [100] gives the threshold $L_T/R = 0.375$ which has a larger deviation from the exact kinetic result than the advanced fluid model in [91].

The threshold $L_T/R = 0.350$ was also obtained independently in [103] (Note that [108] originates from 1988). We moreover note that the quasilinear correspondence to the upper stability regime in ε_n is a pinch flux proportional to ε_n as seen in (6.152), (6.153). Here only the div $\mathbf{q}_*$ part contributes, i.e. the part that changes the sign of the wave energy. The stabilizing effects of $\mathbf{q}_*$ are stronger in the hot ion regime. This is true both linearly and for the pinch flux. This is, in fact the main reason for the good confinement in the hot ion regime in this type of theory. We finally note that as a consequence of the upper stability regime in ε_n, the toroidal η_i mode is stable near the axis in tokamaks. This was first pointed out in [96].

6.10.2 Closure

The reason for the truncation of the above advanced fluid model by taking $\mathbf{q} = \mathbf{q}_*$ is that $\mathbf{q}_*$ is the highest moment that depends only on the moments that are normally fed by sources (fuelling, heating) in a magnetic confinement device. In general we have:

$$\langle v_i v_j v_k v_l \rangle = \langle v_i v_j \rangle \langle v_k v_l \rangle + \ldots\ldots + G(\mathbf{r}, t) \tag{6.133}$$

where $G = \langle v_i v_j v_k v_l \rangle_{irr}$ is the irreducible part.

The transport equation for G can be written in the form:

$$\frac{\partial G}{\partial t} = \frac{1}{r}\frac{\partial}{\partial r}\left(r\chi_G \frac{\partial G}{\partial r}\right) + S_G \tag{6.134}$$

The formal procedure in deriving the fluid model is to approximate the four velocity correlation in the heat-flow equation with products of two velocity correlations which means taking $G = 0$. Higher order moments (i.e. G above) will not have sources in their transport equations ($S_G = 0$) and should decay on a timescale of the order of the confinement time while the moments that are fed by sources remain in quasi-stationary states for many confinement times. Thus taking $G = 0$ leads to one diamagnetic heat flow for parallel temperature and one for perpendicular temperature (6.29), (6.30) [137]. For isotropic temperature the Braghinskii $\mathbf{q}_*$ is recovered as the sum of these. As it turns out, the energy equations for parallel and perpendicular temperatures, contain nonlinearities that tend to isotropize the temperature perturbations. Already in the linear regime we conclude that the effect of temperature unisotropy is small in the 2D case since this is the only plausible explanation for the small difference (5%) in the coefficients of the expansion (5.31).

A phenomenon which will be discussed in the following is that of inward "pinch" fluxes. Since these have a tendency to equilibrate length scales (such as $L_T = -T/(\partial T/\partial r))$ we note that a higher moment at a low level becomes very sensitive (as both T and $\partial T/\partial r$ are small) and can easily adjust itself to an equilibration of length scales without affecting the lower moments. An eventual pinch in the transport equation for the higher moment would thus not replace a source. A pinch in the interior actually has to be fed by a source in the outer parts.

A relation that holds for temperatures is that the perturbation of a quantity becomes small if the background of the same quantity is small. Thus extrapolating this relation to the irreducible part of the perturbed four velocity correlation (δ G) we expect it to decay to zero in a confinement time. With this closure, which on timescales longer than the confinement time according to the above arguments will be valid, we can treat the whole range of states from adiabatic to isothermal i.e. with arbitrary relations between frequency and magnetic drift frequency.

In the local limit, the ion density response is now a two-pole response and when parallel ion motion is included it becomes a three-pole response. When we include higher fluid moments, the order of the density response increases by one for each new moment. The fluid resonances become more and more densely packed as we increase the order until they form a continuum in the infinite limit. The product of infinitely many fluid resonances in the denominator leads to a kinetic, dissipative resonance.

$$\frac{\delta n}{n} = \frac{\omega - \omega_{*e} + \dots\dots\dots}{(\omega - \alpha_1 \omega_D)(\omega - \alpha_2 \omega_D)\dots\dots(\omega - k_{||} v_{||})\dots} \frac{e\phi}{T_e} \tag{6.135}$$

The fluid closure used here thus includes the fluid resonances that correspond to moments that have sources in the experiment. These resonances form a part of the kinetic resonance. We thus include the part of the kinetic resonance that corresponds to the moments that are maintained by external sources. This part is then treated self-consistently in the transport calculations.

Another related aspect of the fluid hierarchy is that higher order moments are much more sensitive to lower moments than vice versa. One example of this is the well known feature that heat flows are much more sensitive to the temperature profiles than the temperature profiles are to the heat flow. An experience from dealing with higher order linear moments in the local limit (e.g. from [137]) is that the introduction of a new, higher order moment, leads to a large shift in the dispersion function when the former eigenvalue is used but a small shift in the eigenvalue is sufficient to restore the dispersion function to its previous value (or smaller). Thus the higher order moment is very sensitive to the eigenvalue. This may be due to the fact that new poles are introduced by the higher order moment. Comparisons with kinetic nonlocal theory [121, 166] show that higher order moments here have a larger impact on the eigenvalue.

We also note that the complication of an integral eigenvalue problem in kinetic theory [140, 148] is absent in the fluid theory. The only possible approximation in the fluid theory is associated with the closure. In the nonlocal theory with parallel ion motion, it turns out that the difference in linear threshold between kinetic theory and the reactive fluid model can be rather large [121, 166] when s/q is of order 1.

This means that the properties of the fluid model depend on how this discrepancy, mainly due to linear Landau damping, is treated. Our advanced fluid model just ignores this difference in linear theory and only retains moments that can be treated self-consistently. The closure made thus relies on the decay of the moment G on the transport time scale. With this nonlinear closure, the fluid moments kept do not have to converge towards a smaller influence for higher moments. On the contrary, the highest moment kept, **q.**, is one of the most important parts. Some aspects of the velocity space dynamics, with potential importance to the closure are discussed in the subsection "Nonlinear kinetic fluid equations".

Since the closure described here does not make use of dissipation we will call this type of fluid model a "Reactive fluid model".

6.10.3 Gyro-Landau Fluid Models

Gyro-Landau fluid models is a class of fluid models that takes a radically different point of view on the closure problem from that presented above. This class of models is actually somewhat beside the main scope of the present review and we will here only make a brief survey without claims of completeness.

Development of Gyro-Landau fluid models was initiated by the work by Hammett and Perkins on Landau damping in the fluid equations for the slab η_i mode [116] This work introduces Landau damping through an imaginary parallel heat flow (q) in the energy equation and is able to recover linear kinetic results for the slab η_i mode. A follow up paper discussed the details of the closure and how the result depends on at which level in the fluid hierarchy the dissipation is introduced [117]. Toroidal effects and with them magnetic drift resonances were introduced by Waltz, Dominguez and Hammet [142]. This work also included FLR effects to all orders. The gyro fluid equations were derived by taking moments of the gyrokinetic equation (5.28) and agreement was obtained with the reactive fluid model described in the previous subsection [108] in the appropriate limit. The closure in this Gyro-Landau model can be written:

$$\mathbf{q} = \mathbf{q}_* + i\mathbf{q}_{gl} \tag{6.136}$$

where $\mathbf{q}_{gl}$ represents the contribution to the resonance from infinitely many higher order moments as obtained by a fit to a Maxwellian velocity distribution. Also this model gives very good agreement with linear kinetic theory. Turbulence simulations in three dimensions have been performed with this model [155].

The fundamental assumption in Gyro-Landau models is that the Gyro-Landau resonance, obtained by a fit to linear kinetic theory, can be used in transport models operating in a nonlinearly saturated state. More recent Gyro-Landau fluid models [157, 163] make the closure at a higher level in the fluid hierarchy but the basic principle of the closure is the same. As we will see in the next section, the Hammett Perkins model for Landau damping, which is included in most more general

Gyro-Landau fluid models, has severe problems in a coherent three-wave system. This, seems to be generally agreed on. However, in an experimental situation we have a broadband, incoherent, turbulence and for such a system the opinions differ.

6.10.4 Nonlinear Kinetic Fluid Equations

A more complete approach, which can be seen as intermediate to the usual Gyro-Landau models and kinetic theory, is the analytical solution to the Vlasov equation obtained by Mattor and Parker [164] in slab geometry. Here the closure is nonlinear although the background velocity distribution function still is Maxwellian. Resonant particles are assumed to follow the phase velocity of the waves so that an integration over particle velocities can be replaced by an integration over wave phases. This model preserves time reversibility and can support a type of trapping oscillations where the velocity distribution is fixed but the wave phase velocity oscillates due to a periodic nonlinear frequency shift. In the Mattor Parker work closure was obtained by including a nonlinear frequency shift in the unexpanded Plasma dispersion function. However, Landaudamping can be maintained in an expanded version as seen in (4.32). The simplest form, (4.32b) is displayed below. It is clear here that a nonlinear frequency shift, as added to ω can easily change the sign of the Landau resonance.

$$\gamma = \left(\frac{\pi}{2}\right)^{1/2} \omega_{*e} \frac{\omega - \omega_{*e}}{k_{\|} \mathrm{v}_{te}} e^{-\omega^2/(k_{\|} v_{te})} \tag{4.32b}$$

It leads to a considerably lower time averaged saturation level than the Hammet-Perkins theory and to time reversible oscillations after the nonlinear saturation. The maxima of these oscillations are close to the Hammet-Perkins saturation level so the reason for the difference is, in fact, that the Hammet-Perkins model phase locks at the maxima of the three wave oscillations while the average level in the Mattor Parker model corresponds to an averaging over the trapping oscillations or rather of the growthrate in (4.32b). Such oscillations can also be expected to occur when higher order moments relax to a nonlinear equilibrium state. If we include inertia of the resonant particles, so that they do not follow the wave phases exactly, we would expect additional phase mixing and a relaxation to a stationary state. This state would be the attractor where the average force between resonant particles and waves changes sign. In systems with many waves, we would expect much stronger phase mixing of resonant particle orbits and a more quasilinear behaviour. The question of the coherent state was actually addressed in a follow up paper by Holod et al. [190] where a diffusion damping was introduced in order to represent the effect of the background turbulence. We also note that the Mattor Parker system includes both quadratic mode coupling terms and cubic nonlinear frequency shifts. This is the same situation as for the turbulent state and, due to self interactions, the turbulent state will also include nonlinear frequency shifts. The effect of diffusion damping was to make the system

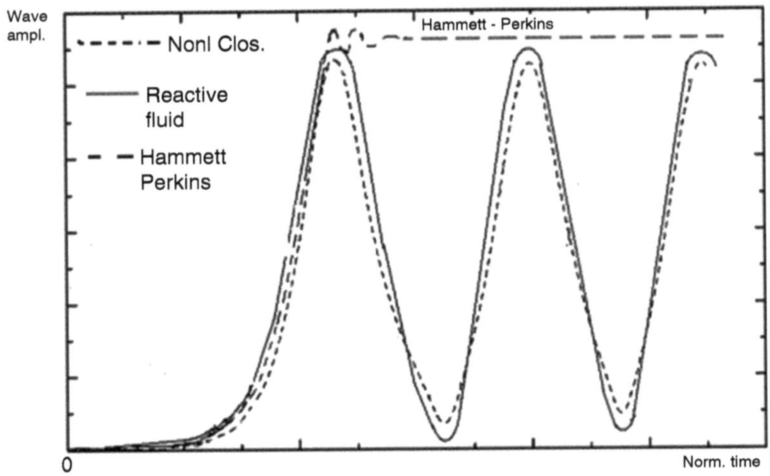

Fig. 6.16 Development in time of three-wave interaction between two slab ITG modes and a zonal flow with different fluid descriptions including reactive fluid, fluid with nonlinear closure and the Hammett Perkins gyro-Landau fluid model (From [197] with permission of the American Institute of Physics)

approach a stationary nonlinear state which appears to be close to the average of the oscillations in the Mattor Parker system. While the Mattor Parker work got good agreement with a fully kinetic system, the work by Holod et al. compared with a reactive closure. A combined qualitative picture is given in Fig. 6.16.

It is important to notice that the kinetic resonance is stabilizing at maxima and destabilizing at minima, thus the effect of the kinetic resonance tends to be averaged out. The work by Holod et al. entered the kinetic integral for the fifth moment while Mattor and Parker used the third moment. However, the reactive ("no closure") result of Holod et al. did not include the fifth moment. It is clear from the small effect of the closure term that the large scale oscillations, in both systems is due to three wave interaction. We also note, that just as in the fully turbulent case there are both quadratic and cubic nonlinearities present. The quadratic nonlinearities phase mix but the cubic do not. The approach to a stationary state is shown by the result from [190] as Fig. 6.17.

We note that the damping due to diffusion leads to an approach to a nonlinear steady state with no energy exchange between particles and waves just as the result of the Fokker-Planck equation in Chap. 9.

6.10.5 Comparisons with Nonlinear Gyrokinetics

Comparisons between Gyro-Landau models and nonlinear gyrokinetics have been going on for several years [154]. Recently the Cyclone group in the US has compared both the magnitude of χ_i and the stiffness (how rapidly χ_i increases

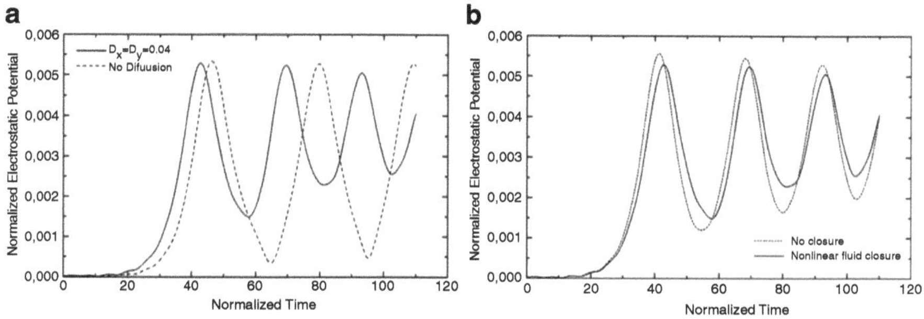

Fig. 6.17 Time evolution of normalized density in a system of three interacting waves with nonlinear closure. Figure (**a**) Compares the case with and without diffusive damping with closure while (**b**) compares the cases with and without closure both with diffusive damping. (From [190] with the permission of AIP)

with the temperature gradient above threshold) of several models [183, 189]. A trend for Gyro-Landau models to give too large transport (up to a factor 3 above the gyrokinetic level) is similar to the results found in [164] where the analytic solution by Mattor and Parker was very close to the full kinetic saturation level while the Hammet-Perkins saturation level was far above. The global and flux-tube gyrokinetic simulations gave somewhat different results in that the flux tube simulations gave more transport and larger stiffness. One of the main questions that have been discussed regarding the saturation level is the damping due to nonlinearly generated background flows [181]. Since these flows have a stronger effect on longer wavelengths, the presence of longer wavelengths in the global simulations may create a more absorbing boundary condition for these as compared to the situation in the flux tube simulations. We note that in this respect the reactive fluid transport model, as described in Sect. 6.11.3, should rather be compared to the global gyrokinetic simulations since an absorbing boundary condition for long wavelengths was used. It is also likely that the mixing length leading to the type (3.67) diffusion coefficient is essential for the stiffness. The scaling $\chi_i \sim \eta_i - \eta_{ith}$, just above threshold, has been seen in mode coupling simulations [93, 100, 105], and was also derived analytically in [130] in the flat density regime. An important point is also that in the comparison by Mattor and Parker, the same three-wave system was used for all models so also the boundary conditions in k-space were the same. This leaves only the different closure schemes as a reason for the differences. On the other hand, since this system contained only three modes, it corresponded to a much more coherent situation than that in the gyrokinetic simulations where a broad spectrum of modes was included. This point was discussed in more detail in the previous section. In the Cyclone simulations also some comparisons with the reactive fluid model discussed above have been made. Preliminary results were reported in [183] and the full results appeared in [189]. An important result was the very large transport obtained by the IFS-PPPL model while our reactive model was in fairly good agreement with the gyrokinetic results (Fig. 6.18).

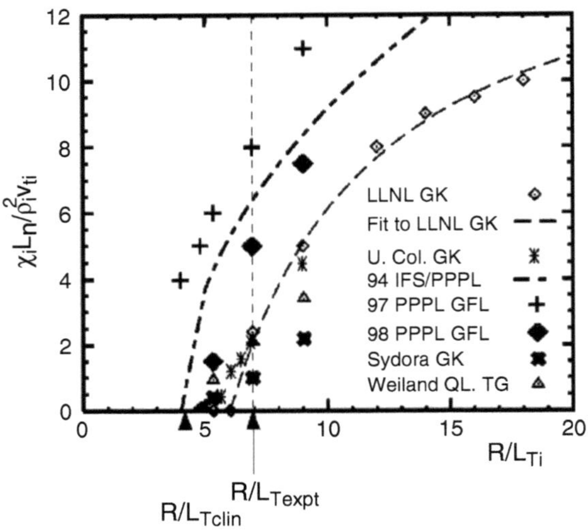

Fig. 6.18 Predicted transport as a function of temperature gradient for different models (From [189] with the permission of the American Institute of Physics)

6.11 Reactive Fluid Model for Strong Curvature

As mentioned several times above a more complete fluid model is needed in order to obtain a correct threshold for η_i, η_e and trapped electron modes. In typical fusion plasmas the magnetic drifts are also comparable to the diamagnetic drifts, except close to the edge. As pointed out above this is the reason for the development of advanced fluid models. A circumstance that improves the possibility for fluid models is that the magnetic drift causes a stream in the plasma. Because of this a fluid resonance, similar to the fluid-beam plasma resonance, is present. Another favourable aspect is that magnetic drifts do not appear explicitly in fluid equations unless the temperature is unisotropic in which case the curvature drift appears. Without magnetic drifts it is clear that the magnetic field localises the particles in the perpendicular direction and that the parameter $k^2\rho^2$ can be chosen as a small parameter to truncate the fluid hierarchy. The usual truncation of the fluid hierarchy is that by Braghinskii. It assumes collision dominance so that the perturbation of the velocity distribution function is Maxwell distributed. In combination with the expansion in $k^2\rho^2$ this leads to the so-called Righi-Leduc or diamagnetic heat flow

$$\mathbf{q} = \mathbf{q}_* = \frac{5}{2}\frac{P}{m\Omega_c}(\widehat{\mathbf{e}}_{\|} \times \nabla T) \tag{6.137}$$

to lowest order in $k^2\rho^2$. In this fluid model the temperature is isotropic. In the beginning of the 1990s a collisionless fluid model was derived by truncating the irreducible part of the fourth moment in the heat-flow equation [137]. In this model no assumptions of Maxwellian distribution was made and temperatures were just defined through quadratic velocity moments. This model gave different **q.** for transport of parallel and perpendicular energy (6.29), (6.30) but when the parallel and perpendicular temperatures were assumed equal the Braghinskii energy equation with the heat flow **q.** given by (6.137) was recovered. Although the temperatures are unisotropic in collisionless linear theory, the isotropic fluid model gives good agreement with Vlasov theory for the toroidal η_i mode concerning threshold and rather good agreement concerning growthrate in the local limit [108]. The main reason for this seems to be that in the low beta case the average of the parallel and perpendicular temperatures enter in the driving pressure term. When parallel ion motion is important, however, unisotropy is essential. We here rely on the nonlinear closure discussed above.

6.11.1 The Toroidal η_i Mode

As discussed above the fluid closure is very important. The energy equation is written

$$\frac{3}{2}n_i\left(\frac{\partial}{\partial t}+\mathbf{v}_i\cdot\nabla\right)T_i+P_i\nabla\cdot\mathbf{v_i}=-\nabla\cdot\mathbf{q}_{*i} \tag{6.138}$$

Where

$$\nabla\cdot\mathbf{q}_{*i}=-\frac{5}{2}n\mathbf{v_{*i}}\cdot\nabla T_i+\frac{5}{2}n\mathbf{v_{Di}}\cdot\nabla T \tag{6.139}$$

Here the first convective diamagnetic part cancels with other convective diamagnetic terms after substitution of the continuity equation for div$\mathbf{v}_i$ as shown in Chap. 2. We will here retain the curvature part of div$\mathbf{q}_{\bullet i}$ which will turn out to be very important. The linearized temperature perturbation is now:

$$\frac{\delta T_i}{T_i}=\frac{\omega}{\omega-5\omega_{Di}/3}\left[\frac{2}{3}\frac{\delta n_i}{n}+\frac{\omega_{*e}}{\omega}\left(\eta_i-\frac{2}{3}\right)\frac{e\phi}{T_e}\right] \tag{6.140}$$

Using (6.140) instead of (6.106), (6.109) is replaced by:

$$\frac{\delta n_i}{n}=\frac{\omega(\omega_{\bullet e}-\omega_{De})+\left(\eta_i-\frac{7}{3}+\frac{5}{3}\varepsilon_n\right)\omega_{*e}\omega_{Di}-k^2{\rho_s}^2(\omega-\omega_{*iT})\left(\omega-\frac{5}{3}\omega_{Di}\right)}{\omega^2-\frac{10}{3}\omega\omega_{Di}+\frac{5}{3}{\omega_{Di}}^2}\frac{e\phi}{T_e} \tag{6.141}$$

where

$$\omega_{*iT} = \omega_{*i}(1. + \eta_i)$$

and we also included the polarisation drift and the lowest order FLR effect, as derived in Chap. 2. The density response (6.141) is of higher order in ω_D than (6.109) both in denominator and numerator. The most important improvement in (6.141) is that it has the right asymptotic limit for large ω_D, i.e. the isothermal limit

$$\frac{\delta n_i}{n} \rightarrow -\frac{e\phi}{T_i} \tag{6.142}$$

for $\omega_D \gg \omega, \omega_{*}$. This can be obtained from (6.109) only in the absence of a temperature gradient. Since δT influences (6.109) only through the temperature gradient one can conclude that a careful treatment of the energy equation is required to make the fluid theory consistent with kinetic theory in the presence of temperature perturbations. The key property of (6.140), absent in (6.106), is that we obtain the correct isothermal limit $\delta T_i \rightarrow 0$ when $\omega_D \gg \omega, \omega_{*}$. This is entirely due to the curvature part of div $\mathbf{q}_{*}$. This part enters as an additional higher order, contribution to the pressure force that may be either destabilizing or stabilizing. The response (6.141) was first applied to MHD ballooning modes [90] where div $\mathbf{q}_{*}$ reproduced an instability below the MHD beta limit previously only seen in kinetic treatments [70]. For ion temperature gradient modes it is usually stabilizing [91]. It is instructive to compare the expansion of (6.141) in ω_D/ω with the corresponding expansion of the gyrokinetic equation, (5.31). These expansions are identical except for the replacement of 7/4 by 5/3. It is fortunate that the terms of order $k^2\rho^2\omega_D/\omega$ agree since no attempt was made to systematically include these in (6.141). The last term prop. to η_i is unsymmetric with respect to $\omega - \omega_{*iT}$. It represents a correction of the basic MHD pressure balance and is, accordingly, responsible for the instability below the MHD beta limit seen in kinetic theory. The fluid model obtained here has sometimes been called fully toroidal since it does not expand in ε_n. Effects of ε_n are, in fact, the most important toroidal effects on drift waves. We emphasize here that the closure (6.137) is treated as exact in the present fluid model. This means that we assume (6.140) and (6.141) to be valid for arbitrary ω_D/ω and these equations should, in general, not be expanded. Clearly the fact that we keep the frequency dependence in (6.140) means that we can describe both slow and fast processes. This increases the number of nonzero poles in the density response (6.141) by one.

By using(6.141) in combination with Boltzmann electrons we obtain the dispersion relation

$$\begin{aligned}
&\omega^2(1 + k^2\rho_s{}^2) - \omega\omega_{*e}\left[1 - \left(1 + \frac{10}{3\tau}\right)\varepsilon_n + k^2\rho_s{}^2\frac{1}{\tau}\left(1 + \eta_i + \frac{5}{3}\varepsilon_n\right)\right] \\
&= \left(\eta_i - \frac{7}{3} + \frac{5}{3}\varepsilon_n\right)\omega_{*i}\omega_{Di} - \frac{5}{3}\omega_{Di}{}^2 - \frac{5}{3}k^2\rho_s{}^2\omega_{*iT}\omega_{Di}
\end{aligned} \tag{6.143}$$

The solution may be written $\omega = \omega_r + i\gamma$ where

$$\omega_r = \frac{1}{2}\omega_{*e}\left[1 - \left(1 + \frac{10}{3\tau}\right)\varepsilon_n - k^2\rho_s{}^2\left(1 + \frac{1+\eta}{\tau} - \varepsilon_n - \frac{5}{3\tau}\varepsilon_n\right)\right] \quad (6.144)$$

and

$$\gamma = \frac{\omega_{*e}\sqrt{\varepsilon_n/\tau}}{1 + k^2\rho_s{}^2}\sqrt{\eta_i - \eta_{ith}} \quad (6.145)$$

Where

$$\eta_{ith} = \frac{2}{3} - \frac{\tau}{2} + \varepsilon_n\left(\frac{\tau}{4} + \frac{10}{9\tau}\right) + \frac{\tau}{4\varepsilon_n} - \frac{k^2\rho_s{}^2}{2\varepsilon_n}$$
$$\times\left[\frac{5}{3} - \frac{\tau}{4} + \frac{\tau}{4\varepsilon_n} - \left(\frac{10}{3} + \frac{\tau}{4} - \frac{10}{9\tau}\right)\varepsilon_n + \left(\frac{5}{3} + \frac{\tau}{4} - \frac{10}{9\tau}\right)\varepsilon_n{}^2\right] \quad (6.146)$$

Here (6.146) was expanded in $k^2\rho_s{}^2$ For very small ε_n terms of the type $k^4\rho_s{}^4/\varepsilon_n$ and with higher powers in ε_n have not been calculated consistently since **q.** has been obtained only to lowest order in $k^2\rho_s{}^2$. They do, however, give the correct trend when compared with kinetic theory. The marginal stability curve in a η_i,ε_n diagram is shown in Fig. 6.15. An important feature as compared to the previous fluid threshold is the presence of an upper stability regime in ε_n. For large ε_n, $\tau = 1$ and $k^2\rho_s{}^2 \to 0$ this threshold is

$$\eta_{ith} = 1.36\varepsilon_n = 2.72L_n/L_B$$

Here L_n scales out and the stability condition becomes

$$L_{Ti} > 0.367L_B \quad (6.147)$$

The correct kinetic threshold here is $0.35L_B$ [108]. The large ε_n regime is often the relevant regime in the bulk of tokamak discharges. This is in particular so for H mode discharges which usually have flat density profiles. The upper stability regime in ε_n also determines how close to the axis the η_i, mode can be unstable, Another interesting aspect of (6.144) is the presence of an $k^4\rho_s{}^4\eta_i{}^2$ term in $\omega_r{}^2$ which enters the stability criterion to the next order in $k^2\rho_s{}^2$. Since this term does not contain ε_n it is in fact consistent and gives an upper stability regime in η_i. This has lead to enhanced confinement states in transport code simulations [114]. In conclusion we thus find that toroidal effects introduce a completely new regime for large ε_n which is dominant in the bulk of tokamaks. The neglect of this regime for a long time caused discrepancies between η_i mode theory and experiments. The new philosophy in the present fluid model, as compared to simple fluid models, is that the closure is made by (6.137) and is assumed to be valid for both slow and fast processes.

6.11.2 Electron Trapping

Since the kinetic integrals for trapped electrons and ions without parallel motion are symmetric [9] (Liu 1969) we may use the same fluid model for the trapped electrons as for the ions. Introducing the fraction of trapped electrons f_t we obtain the dispersion relation [103, 109]:

$$\frac{\omega_{*e}}{N_i}\left[\omega(1-\varepsilon_n)+\left(\eta_i-\frac{7}{3}+\frac{5}{3}\varepsilon_n\right)\omega_{Di}-k^2\rho_s{}^2(\omega-\omega_{*iT})\left(\frac{\omega}{\omega_{*e}}+\frac{5}{3\tau}\varepsilon_n\right)\right]$$
$$=f_t\frac{\omega_{*e}}{N_e}\left[\omega(1-\varepsilon_n)+\left(\eta_e-\frac{7}{3}+\frac{5}{3}\varepsilon_n\right)\omega_{De}\right]+1-f_t \tag{6.148}$$

$$N_j=\omega^2-\frac{10}{3}\omega\omega_{Dj}+\frac{5}{3}\omega_{Dj}{}^2;j=i,\ e \tag{6.149}$$

Here the denominators N_j act as the resonant denominators in the dispersion relation of a two-stream instability. When $N_i < N_e$ the mode propagates in the ion direction (η_i mode) and when $N_e < N_i$ the mode propagates in the electron direction (trapped electron mode). Equation 6.148 is the generalization of (6.126) to arbitrary ε_n and is a quartic equation. Accordingly, it can have two modes unstable at the same time. For ε_n of order 1 the modes are rather independent, propagating in opposite directions, and the dispersion relations can usually be rather well approximated by neglecting the part with the larger N_j in (6.148). For small ε_n, however, the modes are strongly coupled and the directions of propagation may change. For large ε_n and $\eta_i \sim \eta_e$ the η_i mode is the most unstables of the modes. Then ignoring the trapped electron part with denominator N_e we obtain the stability threshold

$$L_{Ti}>\frac{L_B}{\frac{20}{9\tau}(1-f_t)+\frac{\tau}{2(1-f_t)}} \tag{6.150}$$

which is the generalization of (6.147) for finite electron trapping. If we instead take N_i large we obtain a generalization of (6.126) where $\Delta = 0$. This is actually the only way of isolating a trapped electron mode which is driven only by compressibility and electron temperature gradient. This was first done in [102]. Since this mode is obtained for $N_e \ll N_i$ it is due to a fluid resonance. A corresponding mode due to the kinetic resonance was included in [9] and also discussed by Adam et al. [23]. In the same sense also the toroidal η_i mode may be regarded as resonant. It is clear from this discussion that these modes require a description valid for $\omega \sim \omega_D$. The stability boundaries for $\eta_i = \eta_e = \eta$ are shown in Fig. 6.19

The modes present in this system are most clearly shown if we display the growthrates as a function of η for an ε_n where the modes are separated in Fig. 6.20.

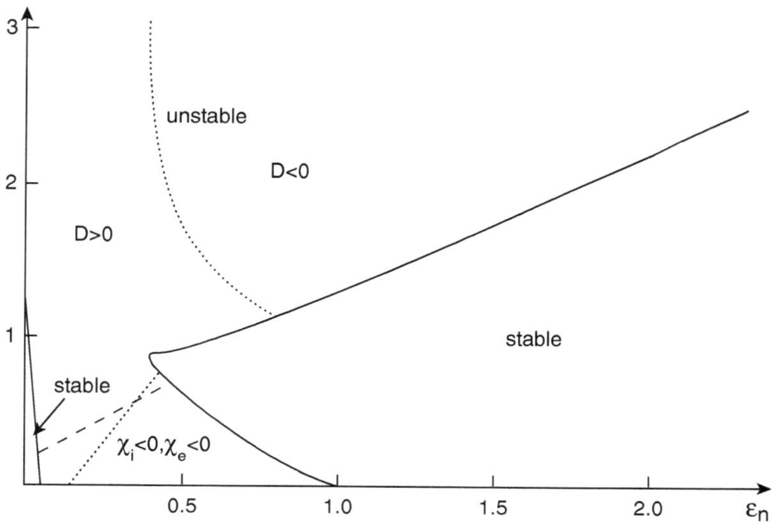

Fig. 6.19 Stability diagram in εn and η (ηi = ηe = η) (From [102] with the courtesy of the IAEA)

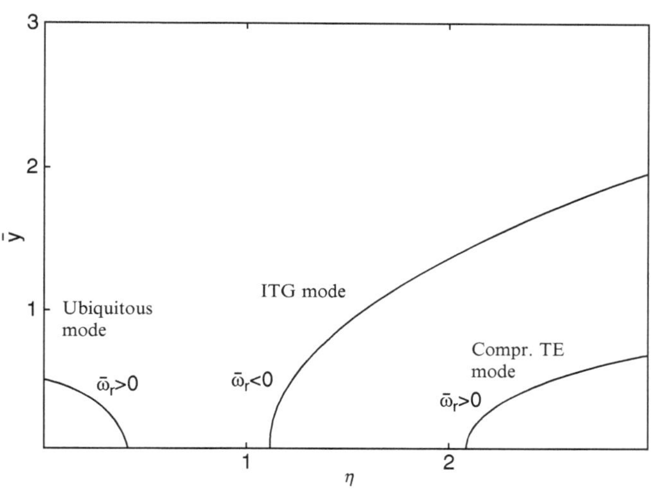

Fig. 6.20 Growth rates as a function of η for εn = 0.8. Other parameters are the same as in Fig 6.19 (From [156] courtesy of the IAEA)

We can see the "Ubiquitous" (trapped electron) mode [19, 28], for small η. The toroidal η_i mode becomes unstable at η just above 1 and then we have the compressional trapped electron mode that becomes unstable at η just above 2 (Fig. 6.20).

The compressional trapped electron mode tends to dominate the transport when $\eta_e > \eta_i$ which is typical when most heating goes to electrons. This is the case for alpha particle heating in burning plasmas. As is shown by Fig. 6.20, the Ubiquitous mode, which is driven by the density gradient (charge separation) is stabilized by the temperature gradient (compressibility). The opposite is true for the ITG mode and the trapped electron mode which is driven by compressibility. This shows also, already at the linear stage, the competition between relaxation of density and temperature inhomogeneities.

6.11.3 *Transport*

We may calculate the quasilinear ion thermal conductivity in the same way as we calculated the particle diffusion in Chap. 2. The thermal conductivity is calculated using Ficks law from

$$\Gamma_T = -\chi \frac{dT}{dr} \tag{6.151}$$

and the saturation level (3.65) can also be derived from the energy equation. The result for the toroidal η_i mode without electron trapping is [93, 100].

$$\chi_i = \frac{1}{\eta_i}\left(\eta_i - \frac{2}{3} - \frac{10}{9\tau}\varepsilon_n\right)\frac{\gamma^3/k_r{}^2}{\left(\omega_r - \frac{5}{3}\omega_{Di}\right)^2 + \gamma^2} \tag{6.152}$$

Equation 6.152 has proven to give good agreement with mode-coupling simulations using many modes if the fastest growing mode ($k^2\rho^2 = 0.1$ in a slab system) is used [93, 101]. The pinch terms (with negative sign) in (6.152) are important. In particular the ε_n pinch term (due to div **q.**) significantly improves the agreement between experimental and theoretical radial profiles of χ_i by suppressing χ_i in the inner region where ε_n is large. The temperature diffusion obtained from (6.152) is always outward since γ is zero if the pinch effects would dominate. If we use the complete fluid model also for the trapped electrons we obtain, as mentioned above, a quartic dispersion relation where both the η_i mode and the trapped electron mode are included and may be unstable simultaneously. In this system there is a coupling between the diffusion of T_i, T_e and n with a trend to equilibrate the equilibrium length-scales L_{Ti}, L_{Te} and L_n. This system contains a possibility of inward fluxes (pinch effects) but the total pressure flux is always outward [102, 109]. It is also interesting to note that in the edge of tokamaks typically $\varepsilon_n \ll 1$, $\eta_i > 1$ and $\gamma > \omega_r$. In this limit (6.152) gives the well known mixing length expression $\chi = \gamma/k_r{}^2$. For the full system with electron trapping the agreement with experimental tokamak transport is remarkably good. In particular the radial profiles of both χ_e and χ_i are usually in rough agreement with experiments at least for $r/a < 0.8$ where a is the small radius, and the magnitude is also usually of the right order.

Furthermore, the ratio χ_i/χ_e is typically about 1/2 at half radius with a radial growth that is somewhat faster than that of f_t. This is also a very typical experimental situation. For the TFTR supershots with peaked density profile and τ about 0.3, χ_e/χ_i is usually small, often less than 1/4 and for the D-III-D hot ion mode with $\tau \sim 0.2$ and flat density profile, large values of χ_e/χ_i are obtained (typically about 4). Both these cases are well reproduced by the fluid model with electron trapping. In self consistent transport code simulations this model also gives the L mode scaling of the energy confinement time with heating power in rough agreement with (1.11) and moreover a spontaneous transition to an H-mode (enhanced confinement regime) with an improvement of τ_E by a factor between 2.5 and 3 for sufficiently strong heating [114]. Here, however, the edge pedestal was not resolved and this result is more an indication that the model supports both L and H mode equilibria in the interior.

The transport coefficients for T_i, T_e and n_e with electron trapping included can be written:

$$\chi_i = \frac{1}{\eta_i}\left(\eta_i - \frac{2}{3} - (1-f_t)\frac{10}{9\tau}\varepsilon_n - \frac{2}{3}f_t\Delta_i\right)\frac{\gamma^3/k_r{}^2}{\left(\omega_r - \frac{5}{3}\omega_{Di}\right)^2 + \gamma^2} \tag{6.153}$$

$$\chi_e = f_t\frac{1}{\eta_e}\left(\eta_e - \frac{2}{3} - \frac{2}{3}f_t\Delta_e\right)\frac{\gamma^3/k_r{}^2}{\left(\omega_r - \frac{5}{3}\omega_{De}\right)^2 + \gamma^2} \tag{6.154}$$

$$D = f_t\Delta_n\frac{\gamma^3/k_r{}^2}{\omega_{*e}{}^2} \tag{6.155}$$

Where, introducing

$$\widehat{\omega} = \omega/\omega_{*e}$$

$$\Delta_i = \frac{1}{N}\left\{|\widehat{\omega}|^2\begin{bmatrix}|\widehat{\omega}|^2(\varepsilon_n - 1) + \widehat{\omega}_r\varepsilon_n\left(\frac{14}{3} - 2\eta_i - \frac{10}{3}\varepsilon_n\right) + \frac{5}{3}\varepsilon_n{}^2\left(-\frac{11}{3} + 2\eta_e + \frac{7}{3}\varepsilon_n\right)\\ -\frac{5}{3\tau}\varepsilon_n{}^2\left(1 + \eta_e - \frac{5}{3}\varepsilon_n\right)\end{bmatrix}\right.$$
$$\left.+\frac{50}{9\tau}\widehat{\omega}_r\varepsilon_n{}^3(1-\varepsilon_n) - \frac{25}{9\tau}\varepsilon_n{}^4\left(\frac{7}{3} - \eta_e - \frac{5}{3}\varepsilon_n\right)\right\} \tag{6.156}$$

$$\Delta_e = \frac{1}{N}\left\{\left|\widehat{\omega}\right|^2\left[\left|\widehat{\omega}\right|^2(\varepsilon_n - 1) + \widehat{\omega}_r\varepsilon_n\left(\frac{14}{3} - 2\eta_e - \frac{10}{3}\varepsilon_n\right) + \frac{5}{3}\varepsilon_n{}^2\left(-\frac{8}{3} + 3\eta_e + \frac{2}{3}\varepsilon_n\right)\right]\right.$$
$$\left.+\frac{50}{9\tau}\widehat{\omega}_r\varepsilon_n{}^3(\varepsilon_n - 1) + \frac{25}{9\tau}\varepsilon_n{}^4\left(\frac{7}{3} - \eta_e - \frac{5}{3}\varepsilon_n\right)\right\} \tag{6.157}$$

$$\Delta_n = \frac{1}{N}\left[\left|\hat{\omega}\right|^2(1-\varepsilon_n) - \hat{\omega}_r\varepsilon_n\left(\frac{14}{3} - 2\eta_i - \frac{10}{3}\varepsilon_n\right) - \frac{5}{3}\varepsilon_n{}^2\left(-\frac{11}{3} + 2\eta_e + \frac{7}{3}\varepsilon_n\right)\right] \tag{6.158}$$

$$N = \left(\hat{\omega}_r{}^2 - \hat{\gamma}^2 - \frac{10}{3}\hat{\omega}_r\varepsilon_n + \frac{5}{3}\varepsilon_n{}^2\right)^2 + 4\hat{\gamma}^2\left(\hat{\omega}_r - \frac{5}{3}\varepsilon_n\right)^2 \tag{6.159}$$

The signs of these in various parameter regimes are indicated in Fig. 6.19.

The most remarkable result obtained with the transport coefficients (6.153)–(6.155) is that from the simulation [129] of the heat pinch on D-III-D [128]. In this simulation, the ECH electron heat source was at half radius and ions were only heated by collisions with electrons. The ECH and Ohmic heating of electrons were taken from the experiment while the particle source at the edge was taken as a free parameter. Both the electron energy pinch and the density and temperature profiles were well reproduced in this simulation. The electron energy pinch was driven by the η_i mode. Also other mechanisms, associated with the toroidal curvature and trapped electrons have been suggested [161].

6.11.4 Normalization of Transport Coefficients

In deriving the transport coefficients (6.153)–(6.155) we used the saturation level (3.65). This saturation level was obtained by balancing linear growth with nonlinear effects at the correlation length scale [93]. Thus the nonlinear effects are here entirely stabilizing. This corresponds to a situation where nonlinear mode coupling carries energy and momentum away from the linearly unstable region in k-space and nothing comes back, i.e. we have absorbing boundaries both at short and long space scales. This saturation level has recently been recovered by a non-Markovian Fokker Planck theory [186]. Both the ion thermal conductivity without trapping, (6.152) and the transport coefficients (6.153)–(6.155) have been normalized and tested against nonlinear mode coupling simulations with absorbing boundaries both at short and long scales [93, 101, 109]. Good agreement was obtained when the FLR parameter $k^2\rho^2$ was about 0.1. This corresponds to the linearly fastest growing mode. Here k should be interpreted as the inverse correlation length. This can be understood by observing that the correlation length typically is determined by the shortest space scales that are strongly excited and this is often given by the source region. It was also verified in mode coupling simulations that the saturation level (3.67) gave better agreement than the mixing length estimate. The damping for long space scales was, in the mode coupling code, obtained from viscosity while that for long wavelengths was artificial. We note, however, that damping due to sheared background flows increases with the space scale and thus has the properties we need for absorbing boundaries at large space scales.

6.11.5 Finite Larmor Radius Stabilization

As seen from (6.143) FLR effects are usually rather marginal for the pure η_i mode. When electron trapping is included it does, however become stronger. A particular limit, in which we can see this explicitely is for $\varepsilon_n \ll 1, \eta_i, \eta_e \gg 1$ so that $\eta \sim 1/\varepsilon_n$. In this limit the dispersion relation splits into two second degree equations, one for $\omega \sim \omega_*$ and one for $\omega \sim \omega_D$. The first case leads to the dispersion relation [114].

$$\widehat{\omega}^2 - \widehat{\omega}\left[1 - \frac{10}{3\tau}\varepsilon_n\left(1 - \frac{1}{\tau}\right) - \frac{k^2\rho_s{}^2}{\tau}\frac{\eta_i}{\Lambda}\right] = -\frac{\varepsilon_n}{\Lambda}\left(\frac{\eta_i}{\tau} + f_t\eta_e\right) \tag{6.160}$$

where $\Lambda = 1 - f_t + k^2\rho_s{}^2$. The growthrate is

$$\widehat{\gamma} = \sqrt{\frac{\varepsilon_n}{\Lambda}\left(\frac{\eta_i}{\tau} + f_t\eta_e\right) - \frac{1}{4}\frac{k^4\rho_s{}^4}{\tau^2}\frac{\eta_i{}^2}{\Lambda^2}} \tag{6.161}$$

We note that the FLR stabilization is fourth order in $k\rho_s$ and corresponds to an upper stability regime in η_i. It is actually a stability regime for steep temperature gradients since L_n can be taken out of (6.161). For the pure mode this regime typically starts at $\eta_i \sim 50$ for $k^2\rho_s{}^2 \approx 0.1$. We note, however, that if $f_t \to 1$, $\Lambda \to k^2\rho_s{}^2$ and the FLR stabilization is only second order in $k\rho_s$ (due to the denominator of the first term). For $f_t \approx 0.6$ a stabilization was obtained in predictive transport simulations for $\eta_i, = 15$. This leads to an enhanced confinement regime with an improvement of a factor 2.5 in the confinement time [114].

In the enhanced confinement state only the mode with $\omega \sim \omega_D$ remains. Its dispersion relation can be written:

$$\widehat{\omega}^2 - \widehat{\omega}\frac{10}{3}\varepsilon_n\,\xi = -\frac{5}{3}\varepsilon_n{}^2\delta \tag{6.162}$$

where

$$\xi = \frac{\eta_i - f_t\eta_e}{\eta_i + \tau f_t\eta_e} \tag{6.163}$$

$$\delta = \frac{\eta_i + f_t\eta_e/\tau}{\eta_i + \tau f_t\eta_e} \tag{6.164}$$

We note that L_n cancels out of both ξ and δ so this mode is a pure magnetic drift mode in a regime where $L_n \ll L_B$. The direction of propagation depends on the sign of ξ and the mode requires div $\mathbf{q}_*$ for instability. This mode always produces a particle pinch as is easily seen from (6.155) . However this is even more obvious from the fact that this is a condensation mode where $\delta p = 0$. Thus a temperature drive corresponds to a positive temperature perturbation and this has to be compensated by a negative density perturbation. (Condensation in the usual sense would mean a positive density perturbation).

6.11.6 The Eigenvalue Problem for Toroidal Drift Waves

We will now also briefly consider the eigenvalue problem of toroidal drift modes. We will limit our study to the ion temperature gradient driven mode (η_i mode) with Boltzmann electrons. The description of this mode is obtained by combining the response (6.141) with the influence of parallel ion motion as described by (6.116). This leads for a parallel wavenumber $k_{\|}$ to the response:

$$\frac{\delta n_i}{n} = \frac{\omega(\omega_{*e} - \omega_{De}) + EIH\omega_{*e}\omega_{Di} + \frac{k_{\|}^2 c_s^2}{\omega}\left[\omega - \frac{5}{3}\omega_{Di} - \omega_{*i}\left(\eta_i - \frac{2}{3}\right)\right] - FL}{\omega^2 - \frac{10}{3}\omega\omega_{Di} + \frac{5}{3}\omega_{Di}^2 - \frac{5}{3\tau}(k_{\|}c_s)^2\left(1 - \frac{\omega_{Di}}{\omega}\right)}\frac{e\phi}{T_e} \tag{6.165}$$

$$EIH = \eta_i - \frac{7}{3} + \frac{5}{3}\varepsilon_n \tag{6.166}$$

$$FL = k^2\rho^2(\omega - \omega_{*ip})\left(\omega - \frac{5}{3}\omega_{Di}\right) \tag{6.167}$$

The parallel wavenumber $k_{\|}$ now becomes an operator i.e.

$$ik_{\|} = \hat{\mathbf{e}}_{\|} \cdot \nabla = \frac{1}{qR}\frac{\partial}{\partial\theta} \tag{6.168}$$

but with a simple transformation we can avoid operating on $\omega_D(\theta)$ with $k_{\|}$. Within the ballooning mode formulation [26, 39], we can interpret θ as an extended poloidal angle θ where the operator $(1/qR)\partial/\partial\theta$ includes both poloidal and radial projections on the parallel direction. The eigenvalue equation can be written in the form

$$\frac{\partial^2\phi}{\partial\theta^2} + h\left\{\left[\hat{\omega} - 1 + k_\perp^2\rho_s^2\left(\hat{\omega} + \frac{1+\eta_i}{\tau}\right)\right]A(\theta) + \varepsilon_n g(\theta)\right\}\phi = 0 \tag{6.169}$$

where

$$h = 4k_\theta^2\rho_s^2\frac{q^2\hat{\omega}}{\varepsilon_n^2} \tag{6.170}$$

$$\widehat{\omega} = \frac{\omega}{\omega_{*e}} \tag{6.171}$$

$$g(\theta) = \cos\theta + s(\theta - \theta_0)\sin\theta \tag{6.172}$$

$$k_\perp{}^2 = k_\theta{}^2(1 + s^2\theta^2) \tag{6.173}$$

and

$$A(\theta) = \frac{\widehat{\omega} + 5\varepsilon_n g(\theta)/3\tau}{F + G\varepsilon_n g(\theta)} \tag{6.174}$$

$$F = \widehat{\omega}\left(1 + \frac{5}{3\tau}\right) + \frac{1}{\tau}\left(\eta_i - \frac{2}{3}\right)$$

$$G = \frac{5}{3\tau}\left(1 + \frac{1}{\tau}\right)\omega$$

The boundary conditions of (6.169) are $\partial\varphi/\partial\theta = 0$ at $\theta = 0$ and $\varphi \to 0$ when $\theta \to \infty$

Here θ_0 is a free parameter which can be choosen to maximize the growthrate. Usually its value is zero but we should keep in mind the possibility of other values. This eigenvalue problem, in general, has to be solved numerically. There exist, however, methods of obtaining approximate solutions in most cases of interest. The most important case where we can obtain an exact analytical solution is the strong ballooning limit where we can take $g(\theta) \approx g(0) = 1$. The applicability of this approximation is wider than we might expect since the geodesic curvature (second part of g) increases when the normal curvature decreases. As it turns out, for $s = 1$, $g(\theta)$ increases slowly with θ up to about $\theta \approx 2.8$. The first zero of $g(\theta)$ for $s = 1$ is just before $\theta = 3$. This picture is changed radically for small and negative shear where the η_i mode is considerably more stable. In the strong ballooning limit (6.169) is of the form

$$\frac{\partial^2\phi}{\partial\theta^2} + (\xi + \delta s^2\theta^2)\phi = 0 \tag{6.175}$$

Equation 6.175 has solutions of the form

$$\phi \propto e^{-\alpha\theta^2}$$

where the θ^2 part gives

$$4\alpha^2 + \delta s^2 = 0$$

or

$$\alpha = \pm \frac{1}{2}\sqrt{-\delta s^2} \tag{6.176}$$

The constant part gives

$$\alpha = \xi/2 \tag{6.177}$$

The formal solution is then

$$\xi = -|s| i\sqrt{\delta} \tag{6.178}$$

Here we have used the fact that δ contains ω^2 and that a positive imaginary part of ω must give a localized mode. The right hand side of (6.178) corresponds to a convective shear damping, similar to that obtained for usual drift waves in (6.40). The right hand side of (6.178) also contains the only effect of the parallel ion dynamics so the condition $\xi = 0$ gives the local dispersion relation with the solution (6.144)–(6.145).

Our nonlocal dispersion relation takes the form:

$$\begin{aligned} &\hat{\omega}^2\left(1 + k_\theta{}^2\rho_s{}^2\right) - \left[1 - \varepsilon_n\left(1 + \frac{10}{3\tau}\right) - i\left(1 + \frac{5}{3\tau}\right)\Theta - k_\theta{}^2\rho_s{}^2\frac{1}{\tau}\left(1 + \eta_i + \frac{5}{3}\varepsilon_n\right)\right]\hat{\omega} + \\ &\qquad + \varepsilon_n\left[\Gamma - \frac{5}{3\tau} + \frac{5}{3\tau^2}(1 + \eta_i)k_\theta{}^2\rho_s{}^2\right] + i\Theta\Gamma = 0 \end{aligned} \tag{6.179}$$

where

$$\Gamma = \frac{1}{\tau}\left(\eta_i - \frac{2}{3}\right) + \frac{5}{3\tau}\varepsilon_n\left(1 + \frac{1}{\tau}\right) \tag{6.180}$$

$$\Theta = \frac{\varepsilon_n|s|}{2q}\sqrt{A(0)\left(1 + \frac{1 + \eta_i}{\tau\hat{\omega}}\right)} \tag{6.181}$$

We note that $\varepsilon_n|s|/2q = L_n/L_s$.

Although (6.169) is rather complicated, the stability threshold takes a very simple form where FLR does not enter [166]. It is (note the difference in the definition of ε_n in [166]):

$$\eta_{ith} = \frac{2}{3} + \frac{10}{9\tau}\varepsilon_n \tag{6.182}$$

The part of the threshold which is linear in ε_n is here entirely due to div **q**.. The part $\tau/(4\varepsilon_n)$ in (6.146) is due to the divergence of the E $\times$ B drift and is removed by the parallel ion motion (there are also other contributions from the divergence of the E $\times$ B drift and div **q.** which cancel in (6.146)). A trend for a higher threshold for the pure toroidal mode was also seen in [74]. We also note that for $k_\perp^2\rho_s^2 = 0$, (6.146) agrees with (6.182) at $\varepsilon_n = 1$.

At this point the boundary curve (6.182) is tangent to the stability boundary in the local limit. The influence of parallel ion motion is thus small for the reactive model in the regime where the strong ballooning approximation applies. Equation 6.182 also exactly defines the threshold in η_i for the first factor in (6.152) to be positive. We note that (6.152) is still valid since we eliminated the ion density response so that $k_\|$ does not enter explicitely. This is also true for the system (6.153)–(6.155), including electron trapping. Due to the ω dependence of θ, (6.181) is now a rather complicated function of ω. However, an analytical solution that works in most cases of interest has been found [191]. This solution still requires iterations since the growthrate depends on the modewidth and the modewidth depends on the eigenvalue.

The success of this solution in combination with flowshear is discussed at the end of Chap. 7. The order of magnitude of $\ominus$ is, usually, well described by the first factor $0.5\varepsilon_n|s|/q$. In a recent work [166] it was found that if the total pressure perturbation is used in (2.47) for the FLR term, (6.182) in fact reduces to only the first factor. That model for FLR does, however, not agree with the kinetic expansion (5.31) for terms of the type $\varepsilon_n k_\perp^2\rho_s^2$. We note also that the slab η_i mode is contained in the present formulation. Because of this the imaginary part is destabilizing for small ε_n, effectively removing the stability regime for small ε_n.

Parallel ion motion also has the effect of reducing the threshold slightly in the flat density regime. This is, however, usually a very small effect and the growthrate is somewhat reduced in the region of instability of the local mode. Another interesting result is that the threshold in η_i reduces to 2/3, independent of ε_n, in the adiabatic limit i.e. when we ignore **q**..

The adiabatic model, accordingly does not produce the upper stability regime in ε_n when parallel ion motion is included. Since, as was pointed out above, the linear kinetic threshold (and, in fact, also the threshold of Gyro-Landau fluid models) changes considerably when parallel ion motion is included, unless s/q is small, we find that our reactive model is less sensitive to parallel ion motion than both the simpler adiabatic model and the kinetic model in linear theory.

6.11.7 Early Tests of the Reactive Fluid Model

The reactive fluid transport model described here has been tested against experiments in more complete versions, i.e. including impurities, collisions on trapped electrons, electromagnetic effects, elongation and Shafranov shifts. The most successful overall results have been obtained by the Multi Mode Model

(MMM) in the US [160, 174, 175, 182], which includes our reactive fluid model in the good confinement region. The MMM uses an artificial (empirical) dependence on elongation which is required for a good overall agreement with the database. The pure reactive fluid model has been found to give very good agreement with JET results [171, 176]. In particular very good agreement has been obtained with high performance shots when finite beta and elongation effects were included. The properties of the reactive model can be characterised as: Thermodynamic properties (power scalings, stiffness): Very good, Geometry properties (magnetic q and shear, elongation scalings): Fairly good magnetic shear scaling while the elongation poses problems. One particular aspect of the thermodynamic properties is that the power scaling has generally been found to be in good agreement with experiment. In particular the scaling $\tau_E \sim P^{-2/3}$ which is often quoted in the literature is the worst case obtained with the reactive fluid model. It is obtained for equal ion and electron heating. The exponent -0.5 has been obtained for mainly ion heating which is in agreement with some experiments using only neutral beam heating as discussed in [114]. The good thermodynamic properties are also emphasized by the good agreement with scans in temperature gradient in the Cyclone tests against nonlinear gyrokinetic simulations [183, 189]. This was made with the basic electrostatic version for the pure η_i mode. The result can be recovered from (6.152) multiplied by 3/2 (energy diffusion), using the local eigenvalue i.e. from the theory in [93]. We note that the background density and temperature profiles were kept fixed in the Cyclone simulations. This is equivalent to applying ideal sources in density and temperature that exactly balance the transport.

The electron trapping effects have also recently been compared with linear kinetic theory [187] and tested against perturbative experiments [188]. More recent tests mainly involve momentum transport and will be discussed in that section. The main reference to the MMM95 is [174]. It used only the strong ballooning approximation and did not include momentum transport. Nevertheless this model was very successful. A more recent model is MMM08 [192] which includes momentum transport and the eigenvalue solution described in [191]. Finally MMMv7.1 has just been presented with improved momentum transport and edge physics (Chapter 7, Ref. 58).

6.12 Electromagnetic Modes in Advanced Fluid Description

We will now generalize the MHD type modes discussed previously to an advanced fluid description. Such a description naturally bridges over to the drift waves in the previous section. In the ideal MHD limit we hardly need the advanced fluid aspects since there the growthrate is much larger than the drift frequencies. However, electromagnetic effects may be important for drift waves. Then, in the edge region with H-mode pedestal there is a mixture of drift waves and MHD type modes where we need an advanced fluid description for a unified description.

A key feature of electromagnetic modes is the parallel electric field. We will here make a fairly general derivation in order to be able to describe also the edge region in a tokamak.

6.12.1 Equations for Free Electrons Including Kink Term

Parallel electron motion gives:

$$\left(\frac{\partial}{\partial t}+i\,\nu_e+\mathbf{v}\cdot\nabla\right)\mathrm{v}_{\|\mathrm{e}}=e\left(\nabla_{\|}\phi+\frac{\partial A_{\|}}{\partial t}-(\mathbf{v}_{*eT}\times\delta\mathbf{B})_{\|}\right)-\frac{1}{n}\nabla_{\|}P \tag{6.183}$$

Where $\|$ indicates the component parallel to the background magnetic field. Now with

$$\delta\mathbf{B}=\nabla\times(A_{\|}\widehat{\mathbf{e}}_{\|})=-\widehat{\mathbf{e}}_{\|}\times\nabla A_{\|}$$

We obtain

$$(\mathbf{v}_{*eT}\times\delta\mathbf{B})_{\|}=-\mathbf{v}_{*\mathbf{eT}}\cdot\nabla A_{\|}$$

Then ignoring electron inertia we obtain

$$\frac{\delta n_e}{n_e}=\frac{e}{T_e}\left(\phi-\frac{\omega-\omega_{*eT}}{k_{\|}}A_{\|}\right)-\frac{\delta T_e}{T_e}+i\frac{\mathrm{v}_{\|\mathrm{e}}}{k_{\|}D_e} \tag{6.184}$$

where

$$D_e=\frac{T_e}{m_e\nu_e}$$

Now using isothermal electrons along a perturbed field line we have

$$(\mathbf{B_0}+\delta\mathbf{B})\cdot\nabla(T_{0\mathrm{e}}+\delta T_e)=0$$

Linearizing we get

$$\delta T_e=\eta_{\mathrm{e}}\frac{\omega_{*e}}{k_{\|}}e\mathrm{A}_{\|}$$

The electron temperature perturbations now cancel. Thus:

$$\frac{\delta n_e}{n_e} = \frac{e}{T_e}\left(\phi - \frac{\omega - \omega_{*e}}{k_{||}} A_{||}\right) + i\frac{\mathrm{v}_{||\mathrm{e}}}{k_{||} D_e} \tag{6.185}$$

The continuity equation of e electrons can be written:

$$\frac{\partial n_e}{\partial t} + \nabla \cdot \left[n_e \left(\mathbf{v_E} + \mathbf{v_{*e}} + \mathrm{v}_{||0} \frac{\delta \mathbf{B}_\perp}{\mathrm{B}} + \mathrm{v}_{||} \widehat{\mathbf{e}}_{||} \right) \right] = 0 \tag{6.186}$$

Where we introduced a background parallel electron velocity due to the plasma current.

Then

$$\frac{\partial n_{ef}}{\partial t} + n_0(\mathbf{v_{*e}} - \mathbf{v_{De}}) \cdot \nabla \frac{e\phi}{T_e} + \mathbf{v_{De}} \cdot \nabla \delta n_{ef} + \frac{n_{ef}}{T_e} \mathbf{v_{De}} \cdot \nabla \delta T_e - \frac{1}{e} \frac{\delta \mathbf{B}_\perp}{\mathrm{B}} \cdot \nabla \mathrm{J}_{||0} + n_0 \mathbf{e}_{||} \cdot \nabla \mathrm{v}_{||} = 0 \tag{6.187}$$

Now

$$\frac{\delta \mathbf{B}_\perp}{\mathrm{B}} \cdot \nabla \mathrm{J}_{||0} = \frac{d\mathrm{J}_{||0}}{\mathrm{dr}} \frac{1}{\mathrm{B}r} \frac{\partial A_{||}}{\partial \theta}$$

Then using also the Ampére law

$$j_{||} = -\frac{1}{\mu_0} \Delta A_{||}$$

We obtain

$$\frac{\partial n_e}{n_0} = \frac{\omega_{*e} - \omega_{De}}{\omega - \omega_{De}} \frac{e\phi}{T_e} - \frac{m}{e\mathrm{B}rn_0(\omega - \omega_{De})} \frac{\partial \mathrm{J}_{||0}}{\partial \mathrm{r}} A_{||} + \eta_e \frac{\omega_{De}\omega_{*e}}{k_{||}(\omega - \omega_{De})} \frac{eA_{||}}{T_e} + \frac{k_{||}\mathrm{v}_{||e}}{\omega - \omega_{De}} \tag{6.188}$$

Then combining (6.185) and (6.188) we obtain

$$\frac{e}{T_e}\left(\phi - \frac{\omega - \omega_{*e}}{k_{||}} A_{||}\right) = \frac{\omega_{*e} - \omega_{De}}{\omega - \omega_{De}} \frac{e\phi}{T_e} + \eta_e \frac{\omega_{*e}\omega_{De}}{k_{||}(\omega - \omega_{De})} \frac{eA_{||}}{T_e} - \frac{m}{e\mathrm{B}rn_0(\omega - \omega_{De})} \frac{\partial \mathrm{J}_{||0}}{\partial \mathrm{r}} A_{||} + \mathrm{v}_{||\mathrm{e}} \left(\frac{k_{||}}{\omega - \omega_{De}} - \frac{i}{k_{||} D_e} \right) \tag{6.189}$$

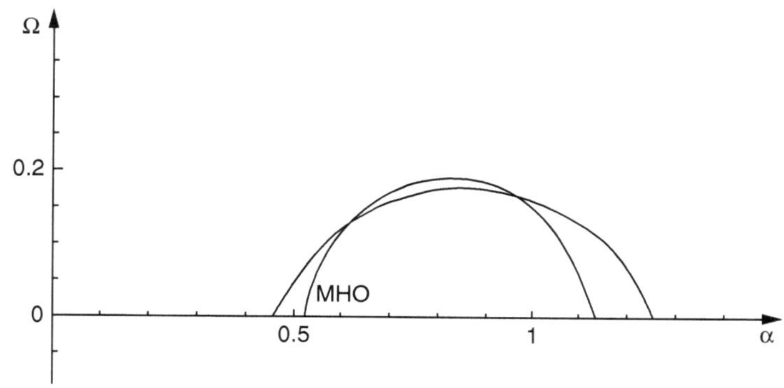

Fig. 6.21 Growthrates of electromagnetic ballooning modes as a function of normalized β. The *inner curve* corresponds to ideal MHD while the *outer* includes Kinetic ballooning modes with a larger unstable region. Here $\varepsilon_n = 0.35$, $\eta i = 2$ and $k2\rho 2 = 0.01$ (From [90] with the permission of the American Institute of Physics)

Now using the Ampére lag again and ignoring parallel ion motion we find

$$\mathrm{v}_{||e} = -\frac{1}{en} j_{||e} \approx -\frac{1}{en} j_{||} = \frac{1}{\mu_0 en} \Delta A_{||} = -k_\perp{}^2 \rho_s{}^2 \mathrm{v_A}{}^2 \frac{eA_{||}}{nT_e}$$

Then (6.189) reduces to

$$\frac{eA_{||}}{T_e} = \frac{k_{||}(\omega - \omega_{*e})}{\omega(\omega - \omega_{*e}) + \omega_{De}(\omega_{*eT} - \omega) - \frac{mk_{||}T_e}{e^2Brn_0}\frac{\partial \mathrm{J}_{||0}}{\partial \mathrm{r}} - \mathrm{k}_\perp{}^2 \rho_s{}^2 k_{||}{}^2 \mathrm{v_A}{}^2 \left(1 - \frac{i(\omega - \omega_{De})}{k_{||}{}^2 D_e}\right)} \frac{e\phi}{T_e} \tag{6.190}$$

Equation (6.190) is our principal result for electromagnetic modes in an advanced fluid description.

We can see that electromagnetic effects vanish when $k_{||}\mathrm{v_A} >> \omega$. We also find that $E_{||} \to 0$when $|\omega| >> |\omega_\bullet|, |\omega_D|$. The relation (6.190), without electron collisions and current has been a standard feature of transport simulations using the reactive drift modes described by (6.153)–(6.155). Recently the full (6.190) has been used in transport code simulations of the H-mode pedestal recovering the peeling mode effects.

6.12.2 *Kinetic Ballooning Modes*

As mentioned previously, the closure (6.137) leads to the inclusion of the kinetic ballooning mode in the electromagnetic case. Again using the ballooning mode formalism but ignoring parallel ion motion we obtain the growth rate of the ballooning mode branch in Fig. 6.21.

As was also pointed out above the kinetic ballooning mode is obtained from the last term in the kinetic expansion (5.31). This is the first term in such an expansion that singles out the temperature gradient from the pressure gradient. Since the same term is due to div **q.** in the fluid equation, it is also a nonadiabatic effect. Although the kinetic ballooning mode is important for limiting the gradient of the H-mode barrier, it is a modest effect as compared to the effect of using only the pressure gradient for toroidal ITG modes. There it gives the η_i threshold -1 as mentioned previously.

6.13 Resistive Edge Modes

The modes considered so far in Chap. 6 have been of a collisionless type which is the most relevant approximation for the core of tokamak plasmas. At the edge, however, the turbulence changes character and collisionless modes are generally not able to explain the continued growth of the transport coefficients outside 80% of the small radius. In the strongly collisional edge ($\nu_{eff} \gg \omega$) we note that (6.101) predicts Boltzmann distributed trapped electrons. This means that trapping is not important. When ν_{eff} becomes larger than the bounce frequency, the trapped electrons behave as free electrons and the most relevant description is to treat all electrons as free and include collisions on them. In a very simple isothermal description ($\delta T_e = 0$) we then arrive at the density response (3.16). Contrary to (6.101) it leads to an MHD type response for large collisionality. This means that collisions prevent the electrons from moving along the field lines. When the electron temperature perturbations are included, the Braghinskii equations lead to an electrostatic parallel electron current of the form:

$$j_{\|e} = enD_e\widehat{\mathbf{e}}_{\|} \cdot \left(\frac{1}{n}\nabla_{\|} n - \frac{e}{T_e}\nabla_{\|}\phi + 1.71\frac{\nabla_{\|}T_e}{T_e}\right) \tag{6.191}$$

where $De = T_e/(0.5m_e\nu_{ei})$. In the electron energy equation we now have to include the contribution from $\kappa_{\|}$ in addition to $\mathbf{q}_{*e}$ in (2.26). The perpendicular collisional heat flow is always smaller than $\mathbf{q}_{*e}$ as long as $\nu_{ei} \ll \Omega_{ce}$. We then obtain:

$$\frac{\delta T_e}{T_e} = \frac{\omega}{\omega - \frac{5}{3}\omega_{D_e} + i\nu_T}\left[\frac{2}{3}\frac{\delta n_e}{n} + \frac{\omega_{*e}}{\omega}\left(\eta_e - \frac{2}{3}\right)\frac{e\phi}{T_e}\right] \tag{6.192}$$

where $\nu_T = 1.06k_{\|}^2D_e$.

In the absence of ν_T, (6.192) is of the same form as (6.140) for ions and exactly the electron temperature perturbation used for collisionless trapped electrons in the derivation of (6.148). In (6.192), however, the two dimensional expression is recovered for very strong collisions since electrons are prevented from moving along the field lines by collisions. We may also point out that fluid closures that make use of kinetic wave-particle resonances can be obtained by a suitable choice of ν_T. In the collisionless regime, we recover the isothermal limit. By using (6.192), (6.191) and the electron continuity equation we can now derive the electron density perturbation in the form

$$\frac{\delta n_e}{n} = \frac{\omega(\omega_{*e} - \omega_{De}) + EEH\omega_{*e}\omega_{De} + ik_{\|}{}^2 D_e T(\omega)}{\omega^2 - \frac{10}{3}\omega\omega_{De} + \frac{5}{3}\omega_{De}{}^2 + ik_{\|}{}^2 D_e \widehat{N}(\omega)} \frac{e\phi}{T_e}$$

$$EEH = \eta_e - \frac{7}{3} + \frac{5}{3}\varepsilon_n \tag{6.193}$$

$$T(\omega) = \omega - \frac{5}{3}\omega_{De} - 1.71\omega_{*e}\left(\eta_e - \frac{2}{3}\right) + 1.06(\omega_{*e} - \omega_{De})$$

$$\widehat{N}(\omega) = \omega - \frac{5}{3}\omega_{De} - 1.14\omega + 1.06(\omega_{*e} - \omega_{De})$$

For ions we use our previous reactive drift wave description i.e. (6.141). In order to obtain an eigenvalue equation we now make the replacement

$$k_{\|}{}^2 \rightarrow -\frac{1}{q^2R^2}\frac{\partial^2}{\partial\theta^2}$$

Since we will here only consider the strong ballooning case, $k_{\|}$ will not operate on $g(\theta)$ and $k_{\perp}(\theta)$ as given by (6.172) and (6.173). we can then obtain our eigenvalue equation directly from (6.193) and (6.141) by letting $k_{\|}$ operate only on φ and we have no problem with non commuting operations. This eigenvalue equation will, however, be of fourth order in general.

Since in the following we will be considering only the strong ballooning approximation we will neglect the fourth order operator, thus arriving at a second order equation of the form:

$$G\frac{D_e}{q^2R^2}\frac{\partial^2\phi}{\partial\theta^2} = -i(A_4\omega^4 + A_3\omega^3 + A_2\omega^2 + A_1\omega + A_0 \tag{6.194}$$

Where

$$G = \left(\omega - \frac{5}{3}\omega_{De}\right)D - [\omega_{*e}(1.71\eta_e - 1.73) - 1.06\omega_{De}]N_i$$
$$- (2.20\omega - 1.06\omega_{De})[\omega\omega_{*e}(1 - \varepsilon_n) + EIH\omega_{*e}\omega_{Di}]$$

where N_i is given by (6.149) and

$$D = \omega^2 - \omega\left[\omega_{*e} - \omega_{De}\left(1 + \frac{10}{3\tau}\right)\right] - EIH\omega_{*e}\omega_{Di} + k_{\perp}{}^2\rho_s{}^2\omega(\omega - \omega_{*iT})$$

is the local dispersion function for the toroidal η_i mode and EIH $= \eta_i - \frac{7}{3} + \frac{5}{3}\varepsilon_n$.
Furthermore

$$\begin{aligned}
A_4 &= k_\perp{}^2\rho_s{}^2 \\
A_3 &= -k_\perp{}^2\rho_s{}^2\left[\omega_{*iT} + \frac{5}{3}\omega_{Di}(1-2\tau)\right] \\
A_2 &= \omega_{*e}\omega_{De}\left\{EEH + \frac{1}{\tau}EIH + \frac{10}{3}(1-\varepsilon_n)\left(1+\frac{1}{\tau}\right)\right. \\
&\left. + \frac{5}{3\tau}k_\perp{}^2\rho_s{}^2\left[\frac{1+\eta_i}{\tau} - 2\left(1+\eta_i+\frac{5}{3}\varepsilon_n\right) + \varepsilon_n\tau\right]\right\} \\
A_1 &= \frac{5}{3}\omega_{*e}\omega_{De}^2\left\{\frac{2}{\tau}(EEH - EIH) - (1-\varepsilon_n)\left(1-\frac{1}{\tau^2}\right)\right. \\
&\left. - k_\perp{}^2\rho_s{}^2\left[\frac{(1+\eta_i)}{\tau}\left(\frac{10}{3\tau}-1\right) - \frac{5}{3\tau}\varepsilon_n\right]\right\} \\
A_0 &= \frac{5}{3}\omega_{*e}\omega_{De}\omega_{Di}\left[\omega_{Di}\left(EEH + \frac{1}{\tau}EIH\right) - \frac{5}{3\tau}k_\perp{}^2\rho_s{}^2\omega_{De}(1+\eta_i)\right]
\end{aligned} \tag{6.195}$$

We note that collisions only enter through D_e in the operator. The local limit of (6.194), corresponding to neglecting the operator is thus identical to the dispersion relation (6.148) in the limit $f_t = 1$. We also note that in the edge region we actually have $\varepsilon_n << 1$ and in connection with strong heating we expect $\eta_i, \eta_e \gg 1$. It is thus interesting to consider the two orderings $\omega \sim \omega_{*e}$ and $\omega \sim \omega_{De}$ discussed in the section 'Finite Larmor radius stabilization'. For large ω we have the resistive ballooning mode.

6.13.1 Resistive Ballooning Modes

Resistive ballooning modes have been studied for a long time both in electromagnetic and electrostatic models. These modes have generally been weakly ballooning with small growth rates. It was recently found [159] that such modes are stable when the shear parameter approaches 1. This effectively rules out this mode as a candidate for edge transport. In the same work, also the presence of a strongly ballooning mode was pointed out. This mode has a growth rate of the ideal MHD order and is thus a very strong candidate for explaining edge transport. The work by Novakovskii et al. [159], however, ignored temperature perturbations which we expect to be very important at the edge in connection with strong heating.

We will here, for simplicity, use the approach of [172] and ignore electron temperature perturbations. This greatly simplifies the algebra at the same time as it retains the important effect of ion temperature gradient on the FLR stabilization (compare (6.161)). Then, including the same geometry as for ion temperature gradient driven modes, (6.172), (6.173) we obtain an eigenvalue equation of the form (6.175). The resulting dispersion relation is

$$\omega(\omega - \omega_{*iT} + i\gamma_D) = \frac{\omega_{*e}\omega_{Di}}{k_\theta^{\,2}\rho_s^{\,2}}(1 + \tau + \eta_i) \tag{6.196}$$

Where

$$\gamma_D = \frac{|s|}{k_\theta \rho_s}\sqrt{-i(\omega - \omega_{*iT})\left(1 - \frac{\omega_{*e}}{\omega}\right)\frac{D_e}{q^2R^2}} \tag{6.197}$$

Here γ_D acts as a shear damping. The right hand side of (6.196) gives an ideal MHD growthrate. When it is fully developed (6.197) can be further simplified. In this limit the exponent α of the eigenfunction as defined in (6.177) can be written

$$\alpha = \frac{|s|}{2}q(k_\theta\rho_s)^{1/2}\sqrt{(\varepsilon_n\Gamma)^{1/2}\frac{\omega_{*e}\nu_{ei}}{v_{the}^{\,2}/R^2}} \tag{6.198}$$

where $\Gamma = 1 + (1 + \eta_i)/2$

We may here realistically estimate the root to be of order 1. For $s \approx 1$ we remember that $g(\theta)$ is close to 1 for $\theta < 3$. The condition for the strong ballooning regime is then $\alpha\theta^2 \geq 1$ i.e. $\alpha \geq 1/9$. This condition is easily fulfilled by (6.198) . In deriving (6.196) we neglected ε_n^2 terms since ε_n is small at the edge and we have been considering frequencies of order ω_{*} or larger. This means that we could have used a simpler fluid model, ignoring div $\mathbf{q}_{*}$. Numerical investigations have shown that the strongly ballooning resistive ballooning mode has its maximum growthrate around

$$k_\theta \rho_s \approx 0.15$$

Below this value the convective damping is the dominant stabilizing mechanism and above this value the FLR stabilization gets more important. In the local limit the condition for FLR stabilization is

$$\frac{1}{4}\omega_{*i}^{\,2}(1 + \eta_i)^2 \geq \frac{\omega_{*i}\omega_{Di}}{k_\theta^{\,2}\rho_s^{\,2}}\tau\Gamma \tag{6.199}$$

With $\varepsilon_n \sim (k_\theta\rho_s)^2$ this condition leads to the stability condition

$$\eta_i \geq 3 \tag{6.200}$$

We note that if electron temperature gradients are included Γ generalizes to $\Gamma = 1 + \eta_e + (1 + \eta_i)/\tau$ which is the combination appearing in the MHD stability parameter α in (6.64). This would lead to a slight increase in the threshold (6.200). It is interesting to compare the threshold (6.200) to that for stabilization due to a poloidal sheared rotation. The neoclassical poloidal rotation v_θ is of the order

$$v_{\theta_i} \sim \eta_i v_{*i} \tag{6.201}$$

A simple version of the Waltz rule gives the threshold

$$\left| \frac{dv_\theta}{dr} \right| \geq \gamma \tag{6.202}$$

where γ is the linear growthrate in the absence of rotation. A natural estimate of dv_θ/dr for steep temperature gradients is

$$\frac{dv_\theta}{dr} \approx \frac{v_\theta}{L_T} \tag{6.203}$$

Then (6.202) leads to the condition

$$\eta_i \geq k_\theta L_{Ti} \sqrt{\frac{\varepsilon_n \tau \Gamma}{(k_\theta \rho_i)^2}} \tag{6.204}$$

Since the root is here typically of order 1, (6.204) expresses the fact that stabilization by rotation at reasonably moderate η requires an L_T of the order of the wave-length of the perturbation. Now, since $L_{Ti} = 0.5\varepsilon_n R/\eta_i$, we obtain for $\varepsilon_n \sim (k_\theta \rho_i)^2$

$$\eta_i \geq (0.5\varepsilon_n k_\theta R)^{2/3} \tag{6.205}$$

In (6.205) we kept only the η_i part of Γ. For typical edge parameters (6.207) gives a threshold $\eta_i \geq 7$. Also this threshold would increase somewhat if we include the electron temperature gradient.

Thus, in conclusion we note that there is a strong ballooning resistive mode for edge parameters with a maximum growthrate of the ideal MHD magnitude. This mode is further stabilized by temperature gradients for $\eta_i \leq 2$–3 and stabilized for $\eta_i \sim 3$–5. This mode is the most likely cause of strong edge transport observed in experiments. We also note that the FLR stabilization of this mode in the local limit is also described by (6.161) for $f_t = 1$. We can thus extend the applicability of (6.161) if we reinterprete f_t as the fraction of electrons that does not move along the magnetic field due to the combined influence of trapping and resistivity. We can, in fact, extend the picture to include also the effect of magnetic induction which also reduces electron motion along the field lines. We may broadly say that (6.160) and (6.161) describe the transition from drift type modes for $f_t = 0$ to MHD type modes for $f_t \rightarrow 1$.

We also note that the FLR stabilization for large η_i typically appears to occur for smaller η_i than the stabilization due to a sheared poloidal rotation. In a scenario when a transition to an enhanced confinement regime is caused by increasing temperature gradients in connection with increasing heating power, we would thus expect the FLR stabilization to set in before the stabilization due to a sheared rotation for neoclassical poloidal rotation. However, as has been found recently, a turbulent spinup of poloidal rotation may change the picture both for the edge barriers and for the internal barriers.

6.13.2 *Transport in the Enhanced Confinement State*

When the resistive ballooning mode described by (6.196) is stable, transport in the system described by (6.193) is strongly reduced, corresponding to an enhanced confinement state. In this regime the relevant ordering of ω is $\omega \sim \omega_D$. With this ordering and $\varepsilon_n \ll 1$, $\varepsilon_n \eta \sim 1$, $\varepsilon_n \sim (k_\theta \rho_i)^2$ and $(k_\perp \rho_s)^2 \sim \varepsilon_n{}^{1/2}$ we obtain an eigenvalue equation which is cubic in ω but which becomes quadratic in the local limit. Again using the geometry defined by (6.172) and (6.173) with solution (6.178) we obtain the dispersion equation:

$$\omega^2 - \frac{10}{3}\xi\omega\omega_{De} + \frac{5}{3}\delta\omega_{De}{}^2 = i\gamma_D{}^2 \tag{6.206}$$

where

$$\xi = \frac{\eta_i - \eta_e + (\tau/2) - (1/2\tau)}{\eta_i + \tau\eta_e + 1 + \tau} \qquad \delta = \frac{\eta_i + (1/\tau)\eta_e - 7(1 + (1/\tau))/3}{\eta_i + \tau\eta_e + 1 + \tau}$$

and

$$\gamma_D{}^2 = k_\theta\rho_s\frac{|s|}{q}\frac{\eta_i}{\eta_i + \tau\eta_e}\sqrt{\frac{5}{3}iFH\frac{D_e}{R^2\omega_{De}}} \tag{6.207}$$

$$H = \left(2 - \frac{1}{\tau}\right)\omega^2 + \left(\frac{10}{3\tau} - 1\right)\omega_{De}\,\omega - \frac{5}{3}\omega_{De}{}^2$$

$$F = \omega_{De}(3.2\omega - 2.7\omega_{De}) - 1.7\tau\frac{\eta_e}{\eta_i}\left(\omega^2 - \frac{10}{3}\omega_{Di}\omega + \frac{5}{3}\omega_{Di}{}^2\right)$$

We have here kept terms of order 1 in the local part since these are actually going to determine the threshold in many cases. In the nonlocal parts we have, however, strictly ignored terms of order 1 as compared to η. The relative complexity of (6.207) is due to the fact that we kept also the electron temperature perturbations and gradients. Here γ_D represents the effects of electron motion along the field lines and the left hand, local part, of (6.206) reduces to (6.162) for $f_t = 1$. We note that for $D_e/R^2 \approx \omega_{De}$, $\gamma_D{}^2$ is typically considerably smaller than ω so that the local approximation is valid. The mode profile is determined by

$$\alpha = \frac{1}{2} k_\theta \rho_s |s| q \sqrt{\frac{5}{3} i \frac{H}{F} \frac{R^2 \omega_{De}}{D_e}} \tag{6.208}$$

Again we note that for $s \approx 1$ we need $\alpha > 1/9$ for the strong ballooning approximation to br valid. This is easy to fulfil for typical edge parameters. The local part of (6.206) has the solution

$$\omega = \frac{5}{3} \xi \omega_{De} \pm \frac{5}{3} \omega_{De} \sqrt{\xi^2 - \frac{3}{5} \delta} \tag{6.209}$$

A necessary condition for instability is clearly

$$\eta_i + \frac{1}{\tau} \eta_e > \frac{7}{3} \left(1 + \frac{1}{\tau} \right) \tag{6.210}$$

$$\frac{\delta T_e}{T_e} = \frac{\eta \omega_{*e}}{\omega - \frac{5}{3} \omega_{De}} \frac{e\phi}{T_e}$$

$$\frac{\delta n_j}{n} = \frac{\eta_j \omega_{*e} \omega_{Dj}}{\omega^2 - \frac{10}{3} \omega_{Dj} \omega + \frac{5}{3} \omega_{Dj}{}^2} \frac{e\varphi}{T_e}$$

This mode is of an MHD character corresponding to $k_\parallel = 0$. It can be obtained from the condition div $\mathbf{j} = 0$ which, when we neglect FLR effects becomes of the form

$$\delta P = 0 \tag{6.211}$$

Here δP is the total perturbed electron plus ion pressure. It means that the density is larger where the temperature is lower and (6.209) is thus a kind of *condensation instability*. This mode is symmetric in ion and electron quantities and has its largest growthrate when $\eta_i = \eta_e$. It cannot be stabilized for large temperature gradients since ξ has η to the same power in numerator and denominator. The reason for this is that the main contributions to the density perturbations come from the convective temperature perturbations through compressibility. This means that a given density perturbation gives rise to a potential perturbation which is inversely proportional to η. This relation

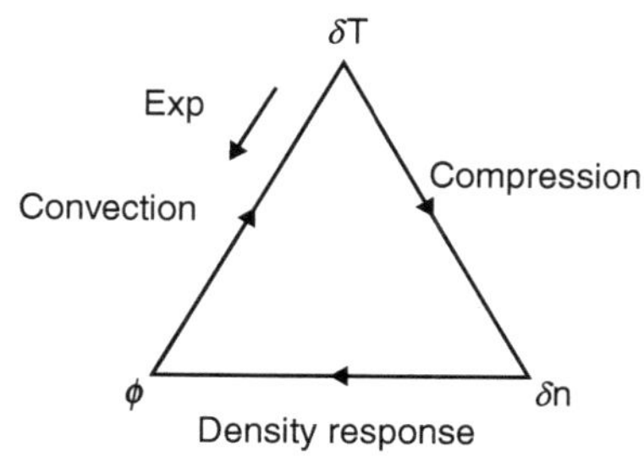

Fig. 6.22 Feedback loop of the condensation instability

replaces the Boltzmann relation in the feedback loop of thermal instabilities (Fig. 6.21). Another consequence of this is that ω would cancel out after taking the final convective temperature perturbation in the feedback loop unless we include the effect of the heat flow. This would remove the phase shift neccessary for instability. *Thus the diamagnetic heat flux is required for instability* (Fig. 6.22).

Finally we repeat what was pointed out in the discussion following (6.151) that this mode always produces a particle pinch. The transport coefficients (6.153)–(6.155) are clearly valid in the local limits for both the resistive ballooning mode and the condensation mode if we take $f_t = 1$. A tendency for a particle pinch at the edge has been seen in several H-mode plasmas [105]. We finally note that an H-mode transition was obtained dynamically in predictive simulations using the transport coefficients(6.153)–(6.155) [114]. This was obtained for $f_t \approx 0.65$ in which case an η_i of 15 was needed for the transition. In our present resistive system we expect the transition to instead occur at $\eta_i \approx 5$.

6.14 Discussion

We have in the present chapter extended the theories of Chaps. 3 and 4 to more realistic geometries. This gave rise to eigenvalue equations that were solved both for some drift type modes and for some MHD type modes. We have also included temperature gradient driven modes. Drift kinetic and gyro-kinetic equations which apply to realistic geometries have been derived and a transport model based on an advanced fluid model for η_i and trapped electron modes has been presented. The general closure problem for fluid models has also been discussed in some detail. Finally we have included a section on resistive edge modes where also a mechanism for the L to H mode transition has been suggested. A condensation mode, able to give a particle pinch in H mode, was also presented. The present chapter essentially shows the present state of research on transport while the MHD parts mainly are included for educational and reference purposes.

Exercises

1. Use the simple trial function $\varphi = 1$ in the quadratic form (6.63) for $\acute{\omega} = 0$. Compare the result with (6.64).
2. Use (6.101) where ν_{eff} is neglected for electrons and include a gravity force for the ions to show that interchange modes can be due to electron trapping.
3. Generalize exercise 4 in Chap. 2 by including curvature effects to a fluid description.
4. Generalize exercise 4 in Chap. 2 by using the drift kinetic equation (5.7), assuming a Maxwell distribution.
5. Derive (6.96).
6. Show that (3.70) is unchanged in the presence of an electron temperature gradient when the equation of state (2.21) is used.

References

1. L.I. Rudakov and R.Z. Sagdeev Sov. Phys. Doklady **6**, 415 (1961).
2. N. A. Krall and M.N. Rosenbluth, Phys. Fluids **5**, 1435 (1962).
3. R.M. Kulsrud, Phys. Fluids **6**, 904 (1963).
4. H. P. Furth, J. Killeen, M.N. Rosenbluth Phys. Fluids **4**, 459 (1963).
5. N.A. Krall andM.N. Rosenbluth Phys. Fluids **8**, 1488 (1965)
6. B. B. Kadomtsev, Plasma Turbulence, Academic Press, New york 1965.
7. B. Coppi, G. Laval, R. Pellat and M.N. Rosenbluth, Nuclear Fusion **6**, 261 (1966).
8. M.N. Rosenbluth, R.Z. Zagdeev, J.B. Taylor and G.M. Zaslavsky, Nuclear Fusion **6**, 297 (1966).
9. C.S. Liu, Phys. Fluids **12**, 1489 (1969).
10. B. B. Kadomtsev and O.P. Pogutse, Soviet Physics Doklady **14**, 470 (1969).
11. P.H. Rutherford and E.A. Frieman, Phys. Fluids **11**, 569 (1968).
12. L.D. Pearlstein and H.L. Berk, Phys. Rev. Lett. **23**, 220 (1969).
13. B.B. Kadomtsev and O.P. Pogutse, in Reviews of Plasma Physics (Ed. M.A. Leontovitch) Consultant Bureau, New York, Vol. 5, p. 249 (1970).
14. B.B. Kadomtsev and O.P. Pogutse, Nuclear Fusion **11**, 67 (1971).
15. M.N. Rosenbluth and M.N. Sloan, Phys. Fluids **14**, 1725 (1971).
16. E.A. Frieman, Phys. Fluids **13**, 490 (1970).
17. P.J. Catto, A.M. El Nadi, C.S. Liu and M.N. Rosenbluth, Nuclear Fusion **14**, 405 (1974).
18. S. Yoshikawa and M. Okabayashi, Phys. Fluids **17**, 1762 (1974).
19. B. Coppi and Rewoldt, Phys. Rev. Lett. **22**, 1329 (1974).
20. D.A. D\'Ippolito and R.C. Davidson, Phys. Fluids **18**, 1507 (1975).
21. V.A. Rozhanskii, JETP Lett. **34**, 56 (1981).
22. M.N. Rosenbluth and W.M. Tang, Phys. Fluids **19**, 1040 (1976).
23. J.C. Adam, W.M. Tang and P.H. Rutherford, Phys. Fluids **19**, 1561 (1976).
24. D.W. Ross, W.M. Tang and J.C. Adam, Phys. Fluids **20**, 613 (1977).
25. T. Hatori, Y.C. Lee and T. Tange, J. Phys. Soc. Japan **43**, 655 (1977).
26. J.B. Taylor in Plasma Physics and Controlled Nuclear Fusion Research IAEA, Vienna 1977, Vol. 2, p. 323.
27. G. Hasselberg, A. Rogister and A. El-Nadi, Phys. Fluids **20**, 982 (1977).
28. B. Coppi and F. Pegoraro, Nuclear Fusion **17**, 969 (1977).
29. B. Coppi, Phys. Rev. Lett. **39**, 939 (1977).

30. D. Dobrott, D.B. Nelson, J.M. Greene, A.H. Glasser, M.S. Chance and E.A. Frieman, Phys. Rev. Lett. **39**, 943 (1977).
31. M. Gaudreau, A., Gondhalekar, M.H. Hughes et. al., Phys. Rev. Lett **39**, 1266 (1977).
32. J.G. Cordey and R.J. Hastie, Nuclear Fusion **17**, 523 (1977).
33. L. Chen, J. Hsu, P.K. Kaw and P.H. Rutherford, Nuclear Fusion **18**, 137 (1978).
34. J.D. Callen, Phys. Rev. Lett. **39**, 1540 (1977).
35. T.M. Antonsen, Jr., Phys. Rev. Lett. **41**, 33 (1978).
36. W.M. Tang, Nuclear Fusion **18**, 1089 (1978).
37. A.B. Mikhailovskii, Sov. J. Plasma Physics **4**, 683 (1978).
38. J.A. Wesson, Nuclear Fusion **18**, 87 (1978).
39. J.W. Connor, R.J. Hastie and J.B. Taylor, Phys. Rev. Lett. **40**, 396 (1978).
40. D. Lortz and J. Nührenberg, Phys. Rev. Lett. **68A**, 49 (1978).
41. A.B. Rechester and M.N. Rosenbluth, Phys. Rev. Lett. **40**, 38 (1978).
42. J.W. Connor, R.J. Hastie and J.B. Taylor, Proc. Royal Soc. London **A365** (1979).
43. B. Coppi, J. Filreis and F. Pegoraro, Ann. Phys. **121**, 1 (1979).
44. B. Coppi, A. Ferreira, J.W.K. Mark and J.J. Ramos, Nuclear Fusion **19**, 715 (1979).
45. C. Mercier in Plasma Physics and Controlled Nuclear Fusion Research (Proc. 7th Int. Conf. Innsbruck 1978) Vol. I IAEA, Vienna p. 701 (1979).
46. H.R. Strauss, Phys. Fluids **22**, 1079 (1979).
47. D.W. Ross, Phys. Fluids **22**, 1215 (1979).
48. D.W. Ross and S.M. Mahajan, Phys. Fluids **22**, 294 (1979).
49. N.T. Gladd and C.S. Liu, Phys. Fluids **22**, 1289 (1979).
50. P.J. Catto, M.N. Rosenbluth and K.T. Tsang, Phys. Fluids **22**, 1284 (1979).
51. C. Chu, M.S. Chu and T. Ohkawa, Phys. Rev. Lett. **41**, 653 (1978).
52. W.M. Manheimer and T.M. Antonsen Jr, Phys. Fluids **22**, 957 (1979).
53. T.M. Antonsen and B. Lane, Phys. Fluids **23**, 1205 (1980).
54. H. Sanuki, T. Watanabe and M. Watanabe, Phys. Fluids **23**, 158 (1980).
55. L. Chen and C.Z. Cheng, Phys. Fluids **23**, 356 (1980).
56. W.M. Tang, J.W. Connor and R.J. Hastie, Nuclear Fusion **20**, 1439 (1980).
57. J.F. Drake, N.T. Gladd, C.S. Liu and C.L. Chang, Phys. Rev. Lett. **44**, 994 (1980).
58. N.T. Gladd, J.F. Drake, C.L. Chang and C.S. Liu, Phys. Fluids **23**, 1182 (1980).
59. M.N. Rosenbluth, Phys. Rev. Lett. **46**, 1525 (1981).
60. J.P. Mondt, Phys. Fluids **24**, 1279 (1981).
61. W. Horton, D.I. Choi and W.M. Tang, Phys. Fluids **24**, 1077 (1981).
62. C.Z. Cheng and K.T. Tsang, Nuclear Fusion **21**, 643 (1981).
63. H.D. Hazeltine, D.A. Hitchcock and S.M. Mahajan, Phys. Fluids **4**, 180 (1981).
64. R.L. Dewar, J. Manickam, R.C. Grimm and M.S. Chance, Nuclear Fusion **21**, 493 (1981).
65. J.M. Greene and M.S. Chance, Nuclear Fusion **21**, 453 (1981).
66. T.M. Antonsen, Jr., A. Fereira and J.J. Ramos, Plasma Physics **24**, 197 (1982).
67. F. Wagner, G. Becker, K. Behringer et. al, Phys. Rev. Lett. **53**, 1453 (1982).
68. J. Weiland and J.P. Mondt, Phys. Rev. Lett. **48**, 23 (1982).
69. K. Itoh, S. Inoue-Itoh, S. Tokuda and T. Tuda, Nuclear Fusion **22**, 1031 (1982).
70. C.Z. Cheng, Phys. Fluids **25**, 1020 (1982).
71. P. Andersson and J. Weiland, Phys. Rev. **A27**, 1556 (1983).
72. M.N. Rosenbluth, S.T. Tsai, J.W. van Dam and M.G. Engquist, Phys. Rev. Lett. **51**, 1967 (1983).
73. G. Hasselberg and A. Rogister, Nuclear Fusion **23**, 1351 (1983).
74. P.N. Guzdar, Liu Chen, W.M. Tang and P.H. Rutherford, Phys. Fluids **26**, 673 (1983).
75. M. Greenwald, D. Gwinn, S. Milora et. al., Phys. Rev. Lett. **53**, 352 (1984).
76. L. Chen, R.B. White and M.N. Rosenbluth, Phys. Rev. Lett. **52**, 1122 (1984).
77. R.B. White, L. Chen, F. Romanelli and R. Hay, Phys. Fluids **18**, 278 (1985).
78. J. Weiland and L. Chen, Phys. Fluids **28**, 1359 (1985).
79. J.P. Freidberg and J.A. Wesson, Nuclear Fusion **25**, 759 (1985).

80. R.C. Grimm, M.S. Chance, A.M.M. Todd, J. Manickam, M. Okabayashi, W.M. Tang, R.L. Dewar, H. Fishman, S.L. Mendelsohn, D.A. Monticello, M.W. Phillips and M. Reusch, Nuclear Fusion **25**, 805 (1985).
81. P. Andersson and J. Weiland, Nuclear Fusion **25**, 1761 (1985).
82. P.C. Liewer, Nuclear Fusion **25**, 543 (1985).
83. R.R. Dominguez and R.R. Moore, Nuclear Fusion **26**, 85 (1986).
84. P. Andersson and J. Weiland, Phys. Fluids **29**, 1744 (1986).
85. A. Rogister, G. Hasselberg, A. Kaleck, A. Boileau, H.W.H. Van Andel and M. von Hellerman, Nuclear Fusion **26**, 797 (1986).
86. F. Romanelli, W.M. Tang and R.B. White, Nuclear Fusion **26**, 1515 (1986).
87. R.R. Dominguez and R.E. Waltz, Nuclear Fusion **27**, 65 (1987).
88. A.A. Thoul, P.L. Similon and R.N. Sudan, Phys. Rev. Lett. **59**, 1448 (1987).
89. W.M. Tang, G. Rewoldt and L. Chen, Phys. Fluids **29**, 3715 (1986).
90. P. Andersson and J. Weiland, Phys. Fluids **31**, 359 (1988).
91. A. Jarmén, P. Andersson and J. Weiland, Nuclear Fusion **27**, 941 (1987).
92. P. Andersson, A fully toroidal fluid analysis of electrostatic ballooning modes including trapped electrons, Preprint CTH-IEFT/PP-1986-17, Chalmers University of Technology (1986).
93. J. Weiland and H. Nordman, Proc. Varenna-Lausanne workshop "Theory of Fusion Plasmas", Chexbres 1988, p. 451 (1988).
94. H. Nordman and J. Weiland, Proc. Varenna-Lausanne workshop "Theory of Fusion Plasmas", Chexbres 1988, p. 459 (1988).
95. J. Weiland, Comments Plasma Phys. Controlled Fusion **12**, 45 (1988).
96. A. Rogister, G. Hasselberg, F. Waelbroeck and J. Weiland, Nuclear Fusion **28**, 1053 (1988).
97. W. Horton, B.G. Hong and W.M. Tang, Phys. Fluids **31**, 2971 (1988).
98. R. Dominguez and R. Waltz, Phys. Fluids **31**, 3147 (1988).
99. B. Coppi, Phys. Lett. **A128**, 193 (1988).
100. F. Romanelli, Phys. Fluids **B1**, 1018 (1989).
101. H. Nordman and J. Weiland, Nuclear Fusion **29**, 251 (1989).
102. J. Weiland, A. Jarmén and H. Nordman, Nuclear Fusion **29**, 1810 (1989).
103. H. Biglari, P.H. Diamond and M.N. Rosenbluth, Phys. Fluids **B1**, 109 (1989).
104. B.G. Hong, W. Horton and D.I. Choi, Phys. Fluids **B1**, 1589 (1989).
105. S. Hamaguchi and W. Horton, Phys. Fluids **2**, 1833 (1990).
106. R.D. Stambaugh, S.M. Wolfe, R.J. Hawryluk et. al., Phys. Fluids **B2**, 2941 (1990).
107. G. Rewoldt and W.M. Tang, Phys. Fluids **B2**, 318 (1990).
108. J. Nilsson, M. Liljeström and J. Weiland, Phys. Fluids **B2**, 2568 (1990); CTH- IEFT/PP-1998-14, Chalmers University of Technology 1988.
109. H. Nordman and J. Weiland and A. Jarmén, Nuclear Fusion **30**, 983 (1990).
110. S.D. Scott, P.H. Diamond, R.J. Fonck et. al., Phys. Rev. Lett. **64**, 531 (1990).
111. F. Romanelli and S. Briguglio, Phys. Fluids **B2**, 754 (1990).
112. P.W. Terry and P.H. Diamond, Phys. Fluids **B2**, 1128 (1990).
113. O.T. Kingsbury and R.E. Waltz, Phys. Fluids **B3**, 3539 (1991).
114. J. Weiland and H. Nordman, Nuclear Fusion **31**, 390 (1991).
115. P.K. Shukla and J. Weiland, Phys. Lett. **A 137**, 132 (1989).
116. G.W. Hammet and F.W. Perkins, Phys. Rev. Lett. **64**, 3019 (1990).
117. G.W. Hammet, W. Dorland and F.W. Perkins, Phys. Fluids **B4**, 2052 (1992).
118. S.C. Cowley, R.M. Kulsrud and R. Sudan, Phys. Fluids **B3**, 2767 (1991).
119. X. Garbet, F. Mourges and A. Samain, Plasma Phys. Control. Fusion **32**, 917 (1990).
120. B.B. Kadomtsev, Nuclear Fusion **31**, 1301 (1991).
121. J. Weiland and A. Hirose, Nuclear Fusion **32**, 151 (1992).
122. J. Weiland, Phys. Fluids **B4**, 1388 (1992).
123. J.P. Christiansen et. al., Plasma Phys. Control. Fusion **34**, 1881 (1992).
124. K. Itoh, S.-I. Itoh and A. Fukuyama, Phys. Rev. Lett. **69**, 1050 (1992).

125. A.J. Wootton, H.Y.W. Tsui and S. Prager, Plasma Phys. Control. Fusion **44**, 2023 (1992).
126. M.A. Dubois, R. Sabot, B. Pegouriee, H-W. Drawn and A. Geraud, Nuclear Fusion **32**, 1935 (1992).
127. A.B. Mikhailowskii, Electromagnetic Instabilities in an Inhomogeneous Plasma, (ed E.W. Laing), Institute of Physics Publishing, Bristol 1992.
128. T.C. Luce, C.C. Petty and J.C.M. de Haas, Phys. Rev. Lett. **68**, 52 (1992).
129. J. Weiland and H. Nordman, Phys. Fluids **B5**, 1669 (1993).
130. A. Heikkilä and J. Weiland, Phys. Fluids **B5**, 2043 (1993).
131. S.C. Guo and F. Romanelli, Phys. Fluids **B5**, 520 (1993).
132. A. Hirose, Phys. Fluids **5**, 230 (1993).
133. H. Nordman and J. Weiland, Phys. Fluids **B5**, 1032 (1993).
134. M. Fröjdh, M. Liljeström and H. Nordman, Nuclear Fusion **32**, 419 (1992).
135. A. Jarmén and M. Fröjdh, Phys. Fluids **B5**, 4015 (1993).
136. A.B. Hassam and Y.C. Lee, Phys. Fluids **27**, 438 (1984).
137. J.P. Mondt and J. Weiland, Phys. Fluids **B3**, 3248 (1991); Phys. Plasmas **1**, 1096 (1994).
138. J. Nilsson and J. Weiland, Nuclear Fusion **34**, 803 (1994).
139. S.C. Guo and F. Romanelli, Phys. Plasmas **1**, 1101 (1994).
140. J.Q. Dong, W. Horton and J.Y. Kim, Phys. Fluids **B4**, 1867 (1992).
141. J.D. Callen, Phys. Fluids **B4**, 2142 (1992).
142. R.E. Waltz, R.R. Dominguez and G.W. Hammett, Phys. Fluids **B4**, 3138 (1992).
143. D.D. Hua, X.Q. Xu and T.K. Fowler, Phys. Fluids **B4**, 3216 (1992).
144. J.W. Connor and H.R. Wilson, Plasma Phys. Control. Fusion **36**, 719 (1994).
145. J. W. Connor, J. B. Taylor and H.R. Wilson, Phys. Rev. Lett. **70**, 1803 (1993).
146. W. M. Tang, G. Rewoldt, Phys. Fluids **5**, 2451 (1993).
147. F. Romanelli and F. Zonca, Phys. Fluids **5**, 4081 (1993).
148. A. Hirose, Phys. Fluids **B5**, 230 (1993).
149. T. Davydova, D. Jovanovic, J. Vranges and J. Weiland, Phys. Plasmas **1**, 809 (1994).
150. A. Hirose, L. Zhang and M. Elia, Phys. Rev. Lett. **72**, 3993 (1994).
151. F. Wagner and U. Stroth, Plasma Phys. Control. Fusion **35**, 1321 (1993).
152. B.A. Carreras, Plasma Phys. Control. Fusion, 1825 (1991).
153. J. Weiland, Current topics in the Physics of Fluids **1**, 439, (1994), Research trends, Trivandrum 1994.
154. S.E. Parker, W. Dorland, R.A. Santoro, M.A. Beer, Q.P. Liu, W.W. Lee and G.W. Hammett, Phys. Plasmas **1**, 1461 (1994).
155. R.E. Waltz, G.D. Kerbel and J. Milovich, Phys. Plasmas **1**, 2229 (1994).
156. J. Nilsson and J. Weiland, Nuclear Fusion **35**, 497 (1995).
157. M. Kotschenreuther, W. Dorland, M.A. Beer and G.W. Hammet, Phys. Plasmas **2**, 2381 (1995).
158. H. Nordman, A. Jarmén, P. Malinov and M. Persson, Phys. Plasmas {\bf 2}, 3440 (1995).
159. S.V. Novakovskii, P.N. Guzdar, J.F. Drake, C.S. Liu and F. Waelbroeck, Phys. Plasmas **2**, 781 (1995).
160. J. Kinsey, G. Bateman, A. Kritz and A. Redd, Phys. Plasmas **3**, 3334 (1996).
161. M.G. Haines, Plasma Phys. Control. Fusion **38**, 897 (1996).
162. M. Persson, J.L.V. Levandovski and H. Nordman, Phys. Plasmas **3**, 3720 (1996).
163. R.E. Waltz, G.M. Staebler, W. Dorland, G.W. Hammett, M. Kotschenreuther and J.A. Konings, Phys. Plasmas **4**, 2482 (1997).
164. Nathan Mattor and Scott E. Parker, Phys. Rev. Lett. **79**, 3419 (1997).
165. H. Nordman, P. Strand, J. Weiland and J.P. Christiansen, Nuclear Fusion **37**, 413 (1997).
166. S.C. Guo and J. Weiland, Nuclear Fusion **37**, 1095 (1997).
167. B.A. Carreras, IEEE Trans. Plasma Sci. **25**, 1281 (1997).
168. J.P. Mondt, Phys. Plasmas **3**, 939 (1996).
169. A. Bondeson, M. Benda, M. Persson and M.S. Chu, Nuclear Fusion **37**, 1419 (1997).

170. A. Zagorodny, J. Weiland and A. Jarm\'{e}n, Comments Plasma Phys. Controlled Fusion **17**, 353 (1997).
171. H. Nordman, P. Strand, J. Weiland and J.P. Christiansen, Nuclear Fusion **37**, 413 (1997).
172. R. Singh, H. Nordman, J. Anderson and J. Weiland, Phys. Plasmas **5**, 3669 (1998).
173. R.E. Waltz, R.L. Dewar and X. Garbet, Phys. Plasmas **5**, 1784 (1998).
174. G. Bateman, A.H. Kritz, J.E. Kinsey, A.J. Redd and J. Weiland, Phys. Plasmas **5**, 1793 (1998).
175. G. Bateman, A.H. Kritz, J.E. Kinsey and A.J. Redd, Phys. Plasmas **5**, 2355 (1998).
176. P.Strand, H. Nordman, J. Weiland and J.P. Christiansen, Nuclear Fusion **38**, 545 (1998).
177. J. Weiland, H. Nordman and R. Singh, Resistive Edge Modes, A scenario for L-H transition due to heat flux, IAEA-CN-69/THP2/09, Yokohama 1998.
178. A. Zagorodny and J. Weiland, Ukrainian J. Physics **43**, 1402 (1998).
179. T.A. Davydova and J. Weiland, Phys. Plasmas **5**, 3089 (1998).
180. J.P. Christiansen and J.G. Cordey, Nuclear Fusion **38**, 1757 (1998).
181. M.N. Rosenbluth and F.L. Hinton, Phys. Rev. Lett. **80**, 724 (1988).
182. D.R. Mikkelsen, G. Bateman, D. Boucher et. al. Proc. Sixteenth IAEA Fusion Energy Conference, Yokohama 1998 (IAEA, Vienna 1999) Paper IAEA-CN-69/ITER P1/08.
183. M.N. Rosenbluth, "Physics fundamentals for ITER", Invited paper, EPS conference Prague 1998, Plasma Phys. and Control. Fusion **41**, A99 (1999).
184. A. Jarmén, P. Malinov and H. Nordman, Plasma Phys. Control. Fusion **40**, 2041 (1998).
185. R. Singh and J. Weiland, Phys. Plasmas **6**, 1397, (1999).
186. A. Zagorodny and J. Weiland, Phys. Plasmas **6**, 2359 (1999).
187. A.J. Redd, A.H. Kritz, G. Bateman, G. Rewoldt, and W.M. Tang, Phys. Plasmas **6**, 1162 (1999).
188. J.E. Kinsey, R.E. Waltz, and J.C. De Boo, Phys. Plasmas **6**, 1865 (1999).
189. A.M. Dimits, G. Bateman, M.A. Beer, B.J. Cohen, W. Dorland, G.W. Hammett. C. Kim, J.E. Kinsey, M. Kotschenreuther, A.H. Kritz, L.L. Lao, J. Mandrekas, W.M. Nevins, S.E. Parker, A.J. Redd, D.E. Shumaker,R. Sydora and J. Weiland, *Comparisons and Physics Basis of Tokamak Transport Models and Turbulence Simulations*, Lawrence Livermore Nat. Lab. Report UCRL-JC-135376 (1999), Phys. Plasmas **7**, 969 (2000).
190. I. Holod, J. Weiland and A. Zagorodny, *Nonlinear Fluid Closure; Three mode slab ion temperature gradient problem with diffusion*, Phys. Plasmas **9**, 1217 (2002).
191. J. Weiland, *Analytical eigenvalue solution for η_i modes of general modewidth*, Phys. Plasmas **11**, 3238 (2004).
192. F.D.Halpern, A. Eriksson, G. Bateman, A.H. Kritz, A. Pankin, C.M. Wolfe and J. Weiland, *Improved model for transport driven by drift modes in tokamaks*, Phys Plasmas **15**, 012304 (2008).
193. Anatol Zagorodny and Jan Weiland, *Non-Markovian Effects in Turbulent Diffusion in Magnetized Plasmas*. Proc. From Leonardo to ITER: Nonlinear and Coherence Aspects, Göteborg, Sweden 18–19 May 2009 p.72.
194. Jan Weiland, Anatoly Zagorodny and Volodymyr Zasenko, *Fluid and Kinetic Modelling on Timescales Longer than the Confinement Time in Bounded Systems*. Proc. From Leonardo to ITER: Nonlinear and Coherence Aspects, Göteborg, Sweden 18–19 May 2009 p.96.
195. A. Zagorodny and J. Weiland, *Closure, at the Irreducible Part of the Fourth Moment for the case of Constant Coefficients in the Fokker-Planck Equation*, Proc. IFP-CNR –Chalmers Workshop on Nonlinear Phenomena in Fusion Plasmas, Villas Monastero, Varenna June 8–10 2011, page 24.
196. Jan. Weiland, and Anatoly Zagorodny, *Fluid Closure, Relation to Particle Pinches, Fluid Resonances*, Proc. IFP-CNR –Chalmers Workshop on Nonlinear Phenomena in Fusion Plasmas, Villas Monastero, Varenna June 8–10 2011, page 33.
197. J. Weiland, A. Zagorodny, V. Zasenko and N. Shatashvili, *On the Physics Description of Fusion Plasmas*. Joint ITER-IAEA-ICTP Advanced Workshop on "Fusion and Plasma Physics" October 3–14 2011. AIP Conf. Proc. *1445*, (2012), I p34, II p54.

Chapter 7
Transport, Overview and Recent Developments

7.1 Stability and Transport

The research areas of stability and transport have developed very strongly during recent years [1–67]. The development in stability and transport up to 2007 has been reviewed in a very comprehensive way by the ITER Expert groups and the ITPA groups [1, 2]. The first paper was accompanied by some pure modeling papers [3, 4]. However several papers on ITER physics basis were published also between these papers [5–7]. Recents areas of strong interest have been momentum transport [21–34, 54–58, 60, 61, 66–67], Impurity transport [20, 43, 44], Finite beta effects [16, 46, 51, 54, 63]. Critical gradient effects and stiffness [45, 55, 56, 66] and particle and heat pinches [8, 10, 16, 36, 44, 52, 53, 57].

We will here focus on momentum transport and the associated barrier formation since this has been the major subject of interest the last years.

7.2 Momentum Transport

Of course there is a major interest in understanding the formation of both Internal Transport Barriers (ITB) and Edge Transport Barriers (ETB). Edge transport barrier just means the barrier associated with the H-mode and this interest thus stems from 1982 (Chap. 6, Ref. 67) while internal barriers were discovered during the 1990s. In Chap. 6 we found some evidence that FLR stabilization may dominate at the edge. This was, however, for neoclassical rotation and now there is experimental evidence that there is a turbulent spinup of rotation for ITB's. Our simulations indicate that this is also a major effect at the edge. A common feature of ETB's and ITB's is that they are formed when the heating is increased. As it turns out, the nonlinear spinup of the poloidal rotation is triggered by the ion temperature length scale,

J. Weiland, *Stability and Transport in Magnetic Confinement Systems*,
Springer Series on Atomic, Optical, and Plasma Physics 71,
DOI 10.1007/978-1-4614-3743-7_7,

$L_{Ti} = -T_i/dT_i/dr$. Thus we need to increase $-dT_i/dr$ without increasing T_i. At the edge this happens naturally just inside the separatrix since the temperature there is kepy low by the large transport in the scrape off layer. This is easily accomplished in a transport code since the outer boundary is kept fixed. However, it is more difficult for the generation of an ITB. In the interior T and dT_i/dr are usually both increased when we increase the heating and it is not clear what happens to L_T. Thus, in practice we need something more to increase $.dT_i/dr$ locally. This can be obtained by e.g. small magnetic shear. The reason why small magnetic shear reduces transport is that the mode profile gets wider for small shear, thus allowing a tendency for the driving curvature term to average out within the mode profile. That this mechanism works in our reactive fluid model is strengthened by the most recent results on stiffness with rotation [66]. We will here reproduce the main features of a recent paper on this.

The poloidal flux of velocity can be written:

$$T_p = \langle v_{Er} v_\theta \rangle = -D_B{}^2 k_r k_\theta \frac{1}{2} \widehat{\phi}^* \left[\widehat{\phi} + \frac{1}{\tau} \widehat{P}_i \right] + c.c \tag{7.1}$$

Here we consider radial flux of poloidal rotation. This is average flux where all velocities are perturbations. However, this flux enters in the transport equation for the background flux. Since the diamagnetic drift does not convect plasma, it is here only the E × B drift that convects. However, the convected poloidal rotation includes both E × B and diamagnetic components. Thus the pressure perturbation and with it the temperature gradient lengthscale enters here. For the toroidal momentum we use

$$\begin{aligned} m_i N_i \left(\frac{\partial}{\partial t} + 2\vec{U}_{Di} \cdot \vec{\nabla} \right) \delta u_{||} &= -m_i N_i \vec{u}_E \cdot \nabla U_{||0} \\ &\quad - \left[\widehat{e}_{||} \cdot \nabla + U_{||0} \frac{m_i \vec{U}_{D_i}}{T_i} \cdot \nabla \right] \\ &\quad \times \left(\delta p_i + eN_i\phi - \frac{\omega + \omega_{*e}(1+\eta_e)/\tau}{k_{||}c} A_{||} \right) \end{aligned} \tag{7.2}$$

Where N is the background density and U denotes unperturbed drifts. This equation was derived from fluid equations, using the stress tensor [33]. It includes also the Coriolis pinch as found in gyrofluid derivations [28–30]. The parallel momentum perturbation is now

$$\begin{aligned} \delta u_{||} &= -\frac{k_\theta D_B}{\omega - 2\omega_{Di}} \frac{dU_{||0}}{dr} + \frac{\langle k_{||} \rangle + \omega_{Di} U_{||0}/(\tau \cdot c_s{}^2)}{\omega - 2\omega_{Di}} \\ &\quad \times \left(\delta p + ne\phi - \frac{\omega + \omega_{*e}(1+\eta_e)/\tau}{k_{||}c} A_{||} \right) / (m_i n_i) \end{aligned} \tag{7.3}$$

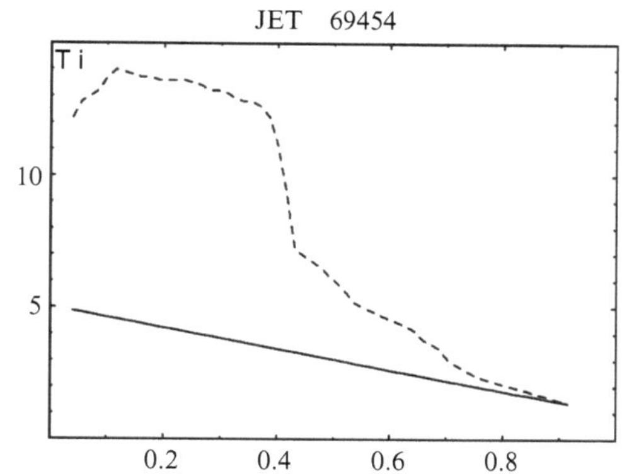

Fig. 7.1 Simulated T_i (*dotted line*) which developed from the initial condition given by the full lune (From [58] with the permission of the American Institute of Physics)

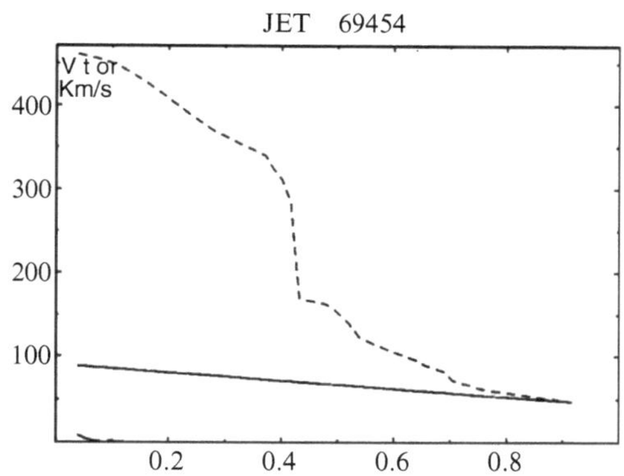

Fig. 7.2 Simulated V_{tor} (*dotted line*) which developed from the initial full (From [58] with permission of the American Institute Physics)

The parallel momentum perturbation was used in order to calculate the toroidal momentum flux in a way analogous to (7.1). The simulated poloidal and toroidal fluxes were then used to calculate the radial electric field as

$$E_r = B_\theta V_\phi - B_\phi V_\theta + \frac{1}{eZ_i n} \frac{\partial P_i}{\partial r} \tag{7.4}$$

7.2.1 *Simulation of an Internal Barrier*

A simulation of JET69454 [58] where the initial conditions had no barrier is shown in Figs. 7.1–7.3. This was a self consistent simultaneous simulation of T_i, T_e V_{tor} and V_{pol}

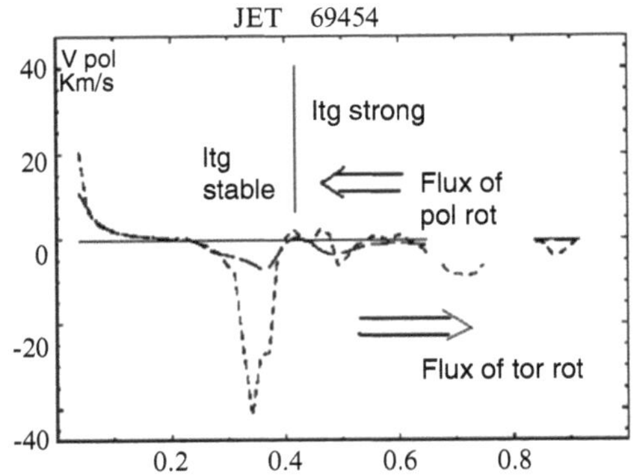

Fig. 7.3 Simulated poloidal spinup (*dotted*), neoclassical rotation (*dashed*), initial condition (full) (From [58] with the permission of the American Institute of Physics)

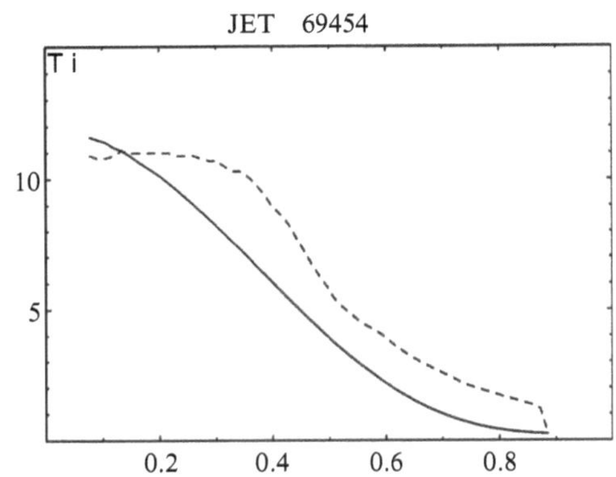

Fig. 7.4 Simulated edge barrier, T_i, in JET69454. (*dotted*) where the *full line* indicates the initial condition (From [58] with the permission of the American Institute of Physics)

7.2.1.1 Poloidal Spinup

The transition requires nonlocal and electromagnetic effects. The pinch of poloidal momentum is driven by the ITG mode and "piles up" at the barrier location where the ITG mode is stable. An electron mode is marginally stable at the barrier.

7.2.2 Simulation of an Edge Barrier

Also the formation of an edge barrier was simulated in [58] Again basic data were taken from JET69454. Here the initial temperatures, including the edge boundary, were reduced by a factor 7 from the experimental condition. As usual in predictive simulations, the edge boundary was then kept fixed. An edge barrier then developed with height approximately equal to the experimental (Figs. 7.4, 7.5, 7.6).

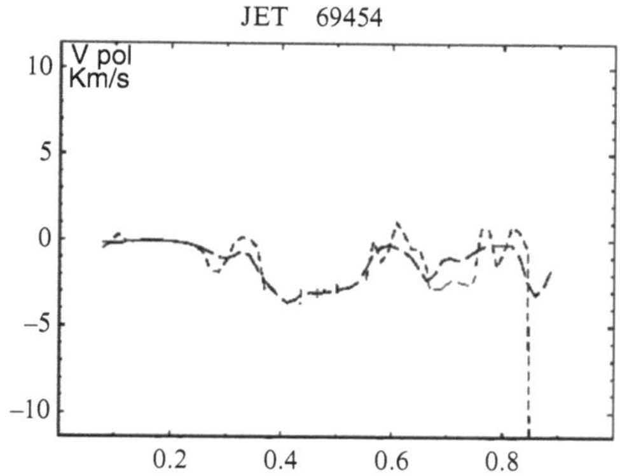

Fig. 7.5 Poloidal rotation in the simulation in 2a (From [58] with the permission of the American Institute of Physics)

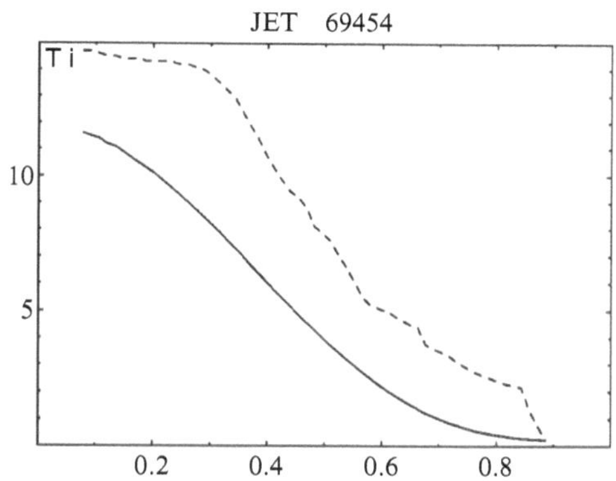

Fig. 7.6 The same shot with a 50% reduction in density and 50% increase in B_p (From [58] with the permission of the American Institute of Physics)

7.2.2.1 Peeling

We then restored the experimental density keeping only the 50% increase in B_p. This activated the kink term and we got peeling (Fig. 7.7).

Both the simulations of the ETB and ITB were made with a code that included the same physics everywhere and used the same grid size everywhere. Thus no information of where the barriers should develop was entered.

We can see here how a fluid model containing both poloidal and toroidal momentum transport can describe the formation of a transport barrier in a selfconsistent simulation of four channels, ion and electron temperature and poloidal and toroidal momenta. The poloidal spinup has previously been recovered also for JET51976 and JET58094 although experimental measurements of the poloidal rotation are missing for 51976. The barrier location is here a result of small magnetic shear (optimized shear) and the power deposition. Rotation is driven by the temperature scale length and this requires both large thermal flux and an additional mechanism which limits transport so that the scale length is

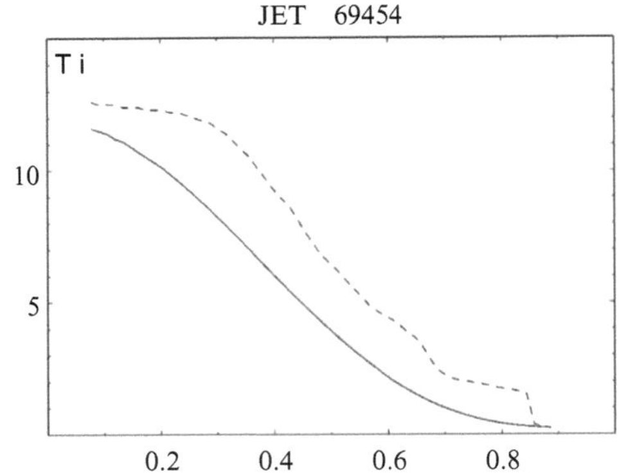

Fig. 7.7 This graph corresponds to Fig 7.4 but with 50% increase in B_p. We observe that the outer part of the barrier has been peeled off (From [58] with the permission of the American Institute of Physics)

reduced. The Trapped Electron mode dominates transport in the whole region of small and negative magnetic shear. However, the ITG mode would also be unstable in the absence of flowshear. Since this model does not include transport due to perturbed magnetic flux surfaces, it is not sensitive to exact values of magnetic q (like rationals). The convergence of the results with respect to resolution has been tested with almost the same results between 50 and 99 radial gridpoints. The results were also tested in the electrostatic limit and in the absence of electron trapping. No internal barrier is formed in these cases. In particular electromagnetic effects have recently been found to be important for the toroidal momentum pinch [54]. This opens up a possible explanation for the stronger barrier in the simulation. The model for elongation is rather crude and usually underestimates the effect. Elongation acts as to reduce electromagnetic effects which, in turn, tend to increase the toroidal momentum pinch. Thus a stronger effect of elongation is expected to reduce the momentum pinch. We have here used a separate correlation length for electron modes using the same method as in [13]. This method has recently been successful in calculating the correlation length in the presence of flowshear [66]. We have made scaling of the edge barrier with edge density and B_p. The height of the barrier is increased when the density is reduced (a factor 0.5 in density increases the edge barrier height by about 15%) and also increased when B_p is increased but this effect is even less sensitive. Peeling is also more effective for large B_p. This is natural since B_p is directly linked to the background current. The transport of toroidal momentum supports, and is actually required for the ITB while the ETB is counteracted. This is because the toroidal momentum flux usually is inward, thus building up rotation in the core and reducing it at the edge, As was found in [54] electromagnetic effects enhance the toroidal momentum pinch and are actually necessary for the formation of the ITB described in this chapter. The dynamics of the ITB formation is that the toroidal momentum pinch builds up rotation in the core and this reduces transport thus increasing the ion temperature gradient. As shown by (7.1) this increases the poloidal rotation and finally the poloidal spinup of poloidal rotation gives the ITB.

7.3 Discussion

We have, in the present chapter given a brief overview of present trends in transport research with emphasis on plasma flows and transport barriers. The two areas of main emphasis in present day research are transport barriers and impurity transport. Impurity diffusion will be of a major importance for ITER. It is essential here that there is a turbulent pinch for the main ions while impurities may go inward or outward. The main ion pinch is likely to improve ITER performance strongly [43, 44]. It is also of a general interest to understand particle and heat pinches [52, 53, 57]. Very basic aspects of a particle pinch in the Levitated Dipole at MIT were studied experimentally in [52] and interpreted theoretically in relation to the particle pinch in the reactive fluid model described in Chap. 6 in [53]. Also the off axis ECH experiment in D-III-D [6.128] was repeated in H-mode [57]. While the experiment in L-mode gave an electron heat pinch as simulated in [6.129], the experiment in H-mode did not. Although we have not simulated the H-mode case it is clear from the start that this would not give a heat pinch since the heat pinch requires a peaked density profile.

Finally the area of stiffness has been studied widely. We will here just discuss the most recent results on stiffness in the presence of rotation. Experiments on stiffness in rotating plasmas has been made for several years [55, 56]. Initially theory models [35] with a spectrum of modes were performing better in relation to the experimental results than the reactive model (Sect. 6.11) with only one correlation length. In this model the spectrum is simulated by using a parameter dependent correlation length [13]. However this correlation length did not depend on flowshear. This has recently been generalized [66]. The result was that the results with this model are now comparable with those of models using a multi mode spectrum [66].

References

1. ITER Physics Basis Editors, ITER, Physics Expert Groups, *ITER PHYSICS BASIS*, Nuclear Fusion **39**, 2137–2664 (9 chapters) Dec. 1999.
2. ITER Physics Basis Editors et. al., *Progress in ITER Physics Basis,* Nucl. Fusion **47**, S1 – S413, (9 chapters) July 2007
3. The ITER 1D Modelling Group, D. Boucher, J.W. Connor, W.A. Houlberg,, M.F. Turner, G. Bracco, A. Chudnovskiy, J.G. Cordey,, M.J Greenwald, G.T. Hoang, G.M.D. Hogeweij, S.M. Kaye, J.E. Kinsey,D.R. Mikkelsen, J. Ongena, D.R. Schissel, H.Shirai, J. Stober, P.M. Stubberfield, R.E. Waltz and J. Weiland, *The International Multi- Tokamak Profile Database,* Nuclear Fusion **40**, 1955 (2000).
4. A.M. Dimits, G. Bateman, M.A. Beer, B.J. Cohen, W. Dorland, G.W. Hammett. C. Kim, J.E. Kinsey, M. Kotschenreuther, A.H. Kritz, L.L. Lao, J. Mandrekas, W.M. Nevins, S.E. Parker, A.J. Redd, D.E. Shumaker,R. Sydora and J. Weiland, *Comparisons and Physics Basis of Tokamak Transport Models and Turbulence Simulations*, Lawrence Livermore Nat. Lab. Report UCRL-JC-135376 (1999), Phys. Plasmas **7**, 969 (2000).

5. V. Mukhovatov, M. Shimada, A. Chudnovsky, A.E. Costley, Y. Gribov, G.Federici, O. Kardaun, A.S. Kukushkin, A. Polevoi, V.D. Pustivitov, Y. Shimomura, T. Sugie, M. Sugihara and G. Vayakis, *Overview of Physics Basis for ITER*, PPCF **45**, A235 (2003).
6. V. Mukhovatov, Y. Shimomura, A. Polevoi, M. Shimada, M. Sugihara, G. Bateman, J.G. Cordey, O. Kardaun, G. Pereverzev, I. Voitsekovich, J. Weiland, O. Zolotukhin, A. Chudnovsky, A.H. Kritz, A. Kukushkin, T.Onjun, A. Pankin and F.W. Perkins, *Comparison of ITER performance predicted by semi-empirical and theory based transport models,* Nucl. Fusion **43,** 942 (2003).
7. M. Shimada, V. Mukhovatov, G. Federici, Y. Gribov, A. Kukushkin, Y. Murakami, A. Polevoi, V. Pustovitov, S. Sengoku and M. Sugihara, *Performance of ITER as a burning plasma experiment*, Nucl. Fusion **44**, 350 (2004).
8. F. Wagner and U. Stroth, *Plasma Phys. Control. Fusion* **35,** 1321-1371 (1993).
9. R. Singh, P.K. Kaw and J. Weiland, Non-Linear features of the electron temperature gradient mode and electron thermal transport in tokamaks. Nuclear Fusion **41**, 1219 (2001).
10. L. Garzotti, X. Garbet, P. Mantica, V. Parail, M. Valovic, G. Corrigan, D. Heading, T.T.C. Jones, P. Lang, H. Nordman, B. Peguorie, G. Saibene, J. Spence, P. Strand, J. Weiland and the EFDA-JET Contributors, "*Particle transport and density profile analysis of different JET plasmas*" Nuclear Fusion **43**, 1829 (2003).
11. J. Pamela J, E.R. Solano, J.M. Adams, et al. *Overview of JET results*, Nuclear Fusion **43**, 1540 (2003).
12. J. Weiland, *Analytical eigenvalue solution for η_i modes of general modewidth*, Phys. Plasmas **11**, 3238 (2004).
13. J. Weiland and I. Holod, *Drift wave transport scalings introduced by varying correlation length*, Phys. Plasmas **12**, 012505-1 (2005).
14. J. Weiland, E. Asp, X. Garbet, P. Mantica, V. Parail, P. Thomas, W. Suttrop and the EFDA-JET contributors, *Effects of temperature ratio on JET transport in hot ion and hot electron* regimes, Plasma Phys. Control. Fusion **47**, 441 (2005).
15. E. Asp, J. Weiland, X. Garbet, P. Mantica, V. Parail, W, Suttrop and the EFDA-JET contributors (2005), *JETTO simulations of T_e/T_i effects on plasma confinement*, Plasma Phys Control. Fusion **47**, 505
16. A. Eriksson and J. Weiland, *Electromagnetic Effects on quasilinear turbulent particle transport,* Physics of Plasmas **12**, 092509-1 (2005).
17. I. Holod, J. Weiland and A. Zagorodny, Nonlinear Fluid Closure; Three mode slab ion temperature gradient problem with diffusion, Phys. Plasmas **9**, 1217 (2002)
18. I. Holod, A. Zagorodny and J. Weiland, *Anisotropic diffusion across an external magnetic field and large-scale fluctuations in magnetized plasmas*, Phys. Rev. **E71**, 046401-1 (2005).
19. P. Mantica, A. Thyagaraya, J. Weiland, G.M.D. Hogeweij and P. J. Knight (2005). *Heat pinches in electron-heated tokamak plasmas: theoretical turbulence models vs. experiments,* Phys. Rev. Letters **95,** 185002-1
20. T. Fülöp and J. Weiland, *Impurity transport in ITER-like plasmas*, Phys. Plasmas **13**, 112504, (2006).
21. J.E. Rice, E.S. Marmar, F. Bombarda and L. Qu, Nuclear Fusion **37**, 421 (1997).
22. X. Garbet, Y. Sarazin, P. Ghendrich, S. Benkadda, P. Beyer, C. Figarella and I Voitsekovitch (2002), Phys. Plasmas **9**, 3893
23. J.S. deGrassie, K.H. Burrell, L.R. Baylor, W. Houlberg and J. Lohr, Phys. Plasmas **11**, 4323 (2004).
24. K. Crombe, Y. Andrew, M. Brix et. al. Phys. Rev. Lett. **95**, 155003 (2005).
25. A.G. Peeters and C. Angioni, Phys. Plasmas **12**, 072515 (2005).
26. P.H. Diamond, S.-I Itoh and T.S. Hahm, Plasma Phys. Control. Fusion **47**, R35 (2005).
27. P.C. deVries, K.M. Rantamäki, C. Giroud, E. Asp, G. Corrigan, A. Eriksson, M. deGreef, I. Jenkins, H.C.M. Knoops, P. Mantica, H. Nordman, P. Strand, T. TALA, J. Weiland, K-D Zastrow and the JET-EFDA contributors, Plasma Phys. Control. Fusion **48**, 1693 (2006).
28. T.S. Hahm, P.H. Diamond, O.D. Gurcan and G. Rewoldt, Phys. Plasmas **14**, 072302-1 (2007).

29. A. G. Peeters, C. Angioni and D. Strinzi, Phys. Rev. Lett. **98**, 265003 (2007).
30. A.G. Peeters, D. Strintzi, Y. Camenen et al, *Influence of the centrifugal force and parallel dynamics on the toroidal momentum transport due to small scale turbulence in tokamaks,* Physics of Plasmas Submitted for publication.
31. D. Strinzi, A.G. Peeters and J. Weiland, Phys. Plasmas **15**, 044502 (2008).
32. T.A. Casper, K.H. Burrell, E.J. Doyle, P. Gohill, C.J. Lasnier, A.W. Leonard, J.M. Moller, T.H. Osborne, P.B. Snyder, D.M. Thomas, J. Weiland and W.P. West, *Density and temperature profile modifications with electron cyclotron power injection in quisent double barrier discharges on DIII-D.* Plasma Phys. Control. Fusion **48**, A35 (2006).
33. J. Weiland, R. Singh, H. Nordman. P.K. Kaw, A. Peeters and D. Strintzi Nuclear Fusion **49**, 065033 (2009).
34. P. Mantica, D. Strintzi, T.Tala et. al. Phys. Rev. Lett. **102**, 175002 (2009).
35. G.M. Staebler et. al. Phys. Plasmas **14**, 055909 (2007).
36. J. Weiland, A. Eriksson, H. Nordman and A. Zagorodny, Plasma Phys. Control. Fusion **49**, A45 (2007).
37. R.E. Waltz, R.R. Dominguez and G.W. Hammett, Phys. Fluids **B4**, 3138 (1992).
38. F.L. Hinton and W. Horton, Phys. Fluids **14**, 116 (1971).
39. A. Rogister, Phys. Plasmas **7**, 5070 (2000).
40. J.J. Ramos, Phys Plasmas **12**, 0512102-1 (2005)
41. J.J. Ramos, Phys. Plasmas **12**, 112301-1 (2005)
42. H.A. Classen, H. Gerhauser, A. Rogister and C. Yarim, Phys. Plasmas **7**, 3699 (2000).
43. J. Weiland, *Predictive Transport Simulations of ITER-FEAT Performance*, Proc 28th EPS Conference on CONTR. FUSION and Plasma Phys. Funchal, Madeira 18 – 22 June 2001. ECA Vol **25A** 633 – 636 (2001).
44. Jan Weiland, Hans Nordman and Pär Strand (2005), *Progress in the theory of anomalous transport in tokamaks, drift waves and nonlinear structuires*, Recent Res. Devel. Physics, **6**, 387
45. E. Asp, J. Weiland, X. Garbet, V. Parail, P. Strand and the JET-EFDA contributors, *Critical Gradient Response of the Weiland Model,* Plasma Phys. Control. Fusion **49**, 1221 (2007).
46. C.E. Kessel, G. Giruzzi,A.C.C. Sips et.al., *Simulation of the hybrid and steady state advanced* operating *modes in ITER*, Nucl. Fusion **47**, 1274 (2007).
47. F.D.Halpern, A. Eriksson, G. Bateman, A.H. Kritz, A. Pankin, C.M. Wolfe and J. Weiland, *Improved model for transport driven by drift modes in tokamaks*, Phys Plasmas **15**, 012304 (2008).
48. V.I Zasenko, A.G. Zagorodny and J. Weiland (2008), *Particle Trapping and Non-Resonant Interaction in a Problem of Stochastic Acceleration*, Ukrainian Journal of Physics **53**, 517
49. V.I Zasenko, A.G. Zagorodny and J. Weiland (2008), *Particle Trapping and Non-Resonant Interaction in a Problem of Stochastic Acceleration*, Ukrainian Journal of Physics **53**, 517
50. A. Zagorodny and J. Weiland, *Non-Markovian renormalization of kinetic coefficients for drift type turbulence in magnetized plasmas. Physics of Plasmas.* **16** 052308. (2009).
51. L. Laborde, D.C. McDonald and I. Voitsekhovitch, *Simulation of Joint European Torus beta scan experiments using theory-based models,* Physics of Plasmas **15,** 102507 (2008).
52. A.C. Boxer, R. Bergmann, J.L. Ellisworth, D.T. Garnier, J. Kesner, M.E. Mauel and P. Woskov, *Turbulent inward pinch of plasma confined by a levitated dipole magnet.* Nature Physics **6**, 207 (2010).
53. J. Weiland, *Turbulence at a pinch,* Nature Physics **6**, 167 (2010).
54. M. Ansar Mahmood, Annika Eriksson and Jan Weiland, *Electromagnetic effects on toroidal momentum transport,* Physics of Plasmas **17**, 122310 (2010).
55. P. Mantica, T. Tala, J. Ferreira, et.al., *Perturbative studies of toroidal momentum transport using neutral beam injection modulation in the Joint European Torus: Experimental results, analysis methodology, and first principles modelling.* Physics of Plasmas. **17,** 092505 (2010).
56. P. Mantica, C. Angioni, C. Challis et. al. Phys. Rev. Lett. **107**, 135004 (2011).

57. M.E. Austin, K.W. Gentle, C.C. Petty, T.L. Rhodes, L. Schmitz and G. Wang, *Heat Transport in Off Axis EC-Heated Discharges in D-III-D*. 51-st APS Meeting of the Division of Plasma Physics, Atlanta, Georgia November 2-6 2009.
58. J. Weiland, K.Krombe, P. Mantica, V. Naulin and T. Tala, *Comparison of Edge and Internal Transport Barriers in Drift Wave Predictive Simulations*, Proc. IFP-CNR –Chalmers Workshop on Nonlinear Phenomena in Fusion Plasmas, Villa Monastero, Varenna June 8 – 10 2011, page 85.
59. T. Rafiq, A.H. Kritz, C. Kessel, G. Bateman, D.C. McCune, R.V. Budny and A.Y Pankin, *Study of Heating and Fusion Power Production in ITER Discharges*. Proc. IFP-CNR –Chalmers Workshop on Nonlinear Phenomena in Fusion Plasmas, Villa Monastero, Varenna June 8 – 10 2011, page 92.
60. Y. Pankin, J.D. Callen, J.R. Cary, R.J. Groebner, A. Hakim, S.E. Kruger, A. Pletzer, S. Shasharina, S. Vadlamani, R.H. Cohen, A.H. Kritz, T.D. Rognlien and T. Rafiq, *Stress Tests of Transport Models Using FACETS Code*, Proc. IFP-CNR –Chalmers Workshop on Nonlinear Phenomena in Fusion Plasmas, Villa Monastero, Varenna June 8 – 10 2011, page 110.
61. N.L. Shatashvili and Z, Yoshida,, *Generalized Beltrami field modeling disk-jet system*. Proc. IFP-CNR –Chalmers Workshop on Nonlinear Phenomena in Fusion Plasmas, Villa Monastero, Varenna June 8 – 10 2011, page 73.
62. Akira. Hirose and Jan. Weiland, *Two-fluid Analysis of the Geodesic Acoustic Mode in Tokamaks* . Proc. IFP-CNR –Chalmers Workshop on Nonlinear Phenomena in Fusion Plasmas, Villa Monastero, Varenna June 8 – 10 2011, page 67.
63. Enzo Lazzaro, Luca Comisso and Marco Del Pra,, *Nonlinear and Diamagnetic Effects in a Neoclassical Model of Magnetic Reconnection* . Proc. IFP-CNR –Chalmers Workshop on Nonlinear Phenomena in Fusion Plasmas, Villa Monastero, Varenna June 8 – 10 2011, page 45.
64. A.I. Smolyakov, S. Benkadda, Y. Camenen, C. Bourdelle and X. Garbet, *Parallel Momentum balance and toroidal rotation in a tokamak*. Proc. IFP-CNR –Chalmers Workshop on Nonlinear Phenomena in Fusion Plasmas, Villa Monastero, Varenna June 8 – 10 2011, page 57.
65. Ettore Minardi, *Approach to the Thermodynamics of High Temperature (non-Maxwellian) Plasmas.*. Proc. IFP-CNR –Chalmers Workshop on Nonlinear Phenomena in Fusion Plasmas, Villa Monastero, Varenna June 8 – 10 2011, page 13.
66. J. Weiland, P. Mantica and the JET-EFDA Contributors, *Effects of flow shear on the correlation length of drift wave turbulence*. 38th EPS Conference, Strassbourg, France, June 27 – July1 2011 Paper P5.130.
67. A.G. Peeters et al. Nuclear Fusion *151*, 094027 (1911)

Chapter 8
Instabilities Associated with Fast Particles in Toroidal Confinement Systems

8.1 General Considerations

As mentioned in Sect. 6.11.3, toroidal drift wave transport gives an unfavourable scaling of the energy confinement time with heating power, roughly in agreement with the empirical scaling law (1.11). It is worth observing that this scaling is obtained with a reactive fluid model where only magnetic drift resonances of a fluid type were included. The unfavourable scaling with heating power is due partly to the scaling of transport coefficients with temperature as $T^{-3/2}$ and partly to the threshold behaviour, i.e. $(\eta_i - \eta_{ith})^{1/2}$. These are effects of a pure (ideal) heating on the bulk plasma transport and are thus independent of the heating method. We note the close analogy with Rayleigh Benard convection in usual fluids where the heating itself leads to convective transport.

A different but somewhat similar picture emerges when we consider how the energy is transformed into heat for a particular heating method. This process in general requires the formation of a non-Maxwellian plasma with an energetic particle population before the external energy is transformed into heat. This is regardless of whether the heating is made by neutral beams, radiofrequency waves or alpha particles in a burning plasma. The fast particle population is here either due to injected or created particles or due to wave-particle resonances with an injected wave. In both cases we need kinetic theory to understand the details of the relaxation. The reason for the interest in the energetic particle population is that it may lead to new instabilities which, in turn, may cause a large transport of the energetic particles. This could lead to a situation where these particles may leave the system before depositing their energy to the bulk plasma, thus reducing the efficiency of the heating method and enhancing the unfavourable scaling of confinement time with heating power given by (1.11). Although instabilities caused by fast particles have been observed experimentally [1–5], the most striking example being the "Fishbone instability" in PDX [1], the scaling (1.11) does not seem to depend strongly on the heating method as such. It may, however, depend on in which channel the energy is deposited. (For the driftwave transport coefficients given by (6.152)–(6.154) the

J. Weiland, *Stability and Transport in Magnetic Confinement Systems*, Springer Series on Atomic, Optical, and Plasma Physics 71,
DOI 10.1007/978-1-4614-3743-7_8,

worst case , P to the power −2/3, is obtained for equal electron and ion heating) This indicates that so far, instabilities caused by the energetic particles have not had a strong effect on the overall energy balance. The reason for this seems to be that the anomalous increase in the transport due to fast particles has been fairly modest. As it turns out, the fast particles also have a beneficial effect [6–8], on the bulk transport which may partly compensate an increased transport in the fast particle channel from an overall energy balance point of view. However, since instabilities caused by energetic particles [9–36], are potentially harmful and since the situation may change in large burning plasmas, such as in a reactor, an understanding of the energetic particle physics may be essential.

8.2 The Development of Research

The first theoretical studies of energetic particle effects indicated a possibility for resonance at the Alfvén frequency [9] and the above mentioned stabilizing effect on eigenmodes associated with the bulk plasma [6–8]. This effect is a dilution effect caused by the fact that the fast particles do not take part in the bulk instabilities. This is true for MHD type modes as well as for drift type modes. Later, however, fast particles were found to introduce new modes at the precession frequency of trapped fast particles [12–14]. These modes were basically of an MHD type since one fluid equations could be used to describe the bulk plasma. The fast particles, however, destabilized a new branch at the precession frequency of the fast particles. These types of modes were, in fact, discovered experimentally as the "Fishbone instability" in the PDX experiment in Princeton [1]. While the main fishbone mode was identified as a new branch of the internal kink mode [12], a precursor, with a higher modenumber seemed to be due to an analogous branch of the high n MHD ballooning mode [14]. Since these modes are driven by resonant fast particles we here have sources in velocity space and thus expect to have sources in the transport equations of the type (6.132) for all fluid moments. Experimentally the fishbone oscillations were obtained for nearly perpendicular neutral beam injection. This led to a large trapped population of the fast particles and the instability was entirely due to the magnetic curvature drift resonance of the trapped particles. (The bounce averaged trapped particles rotate "precess" in the toroidal direction due to magnetic curvature. The bounce averaged magnetic drift is called the precession frequency). The threshold in energetic particle beta of the fishbone instability is due to the continuum damping of the MHD type mode. The MHD continuum for cylindrical plasmas was discussed at the end of section on kink modes in Chap. 6. For the MHD ballooning mode the continuum damping corresponds to the $i\acute{\omega}$ term in (6.72). Good agreement between theory and experiment was obtained for both threshold and mode signatures for the fishbone modes. An obvious way to reduce the effect of fishbones was to inject the neutrals more parallel to the magnetic field. This would reduce the fraction of trapped particles to below the threshold set by the continuum damping. This method, however, turned out to be only partly successful and new types of modes were excited also below the continuum threshold of the fishbone instability. For these modes the Landau resonance of the circulating fast particles becomes important. As it turns out, the plasma current may cause a minimum of the Alfvén frequency as a function of radius, thus opening the possibility for modes with a discrete spectrum when the frequency is below the minimum of the Alfvén frequency. These modes were called Global Alfvén Eigenmodes [15, 16], (GAE) since they could extend over the whole cross-section (compare the discussion at the end of the section on kink modes in Chap. 6). Another possible cause of discrete, undamped modes is toroidicity. (In fact, in a toroidal system also GAE modes can be seen as toroidal since the current enters in combination with the parallel operator $k_{\|}$ prop. to 1/R). The main effect of toroidicity is to couple modes with different poloidal modenumbers. This introduces a gap in the Alfvén continuum due to coupling of modes with poloidal modenumbers m and m + 1. This is the Toroidicity induced Alfvén Eigenmode or TAE mode [17, 18]. For a while it was believed that GAE modes and TAE modes had a very small instability threshold set only by electron Landau damping. Later it was, however, found that these modes can couple to the continuum modes by other toroidal coupling possibilities. This usually gives the main stabilizing effect. There are more types of modes of a similar type. We may mention EAE modes caused by the coupling between m and m + 2 modes due to ellipticity, NAE modes which couple m and m + 3 modes due to triangularity and BAE modes which are due to finite beta modifications of the magnetic curvature.

8.3 Dilution Due to Fast Particles

Before going into the new types of instabilities caused by fast particles we shall make some general considerations of multi ion systems where, in particular, the effect of dilution becomes clear. We consider a system of electrons, e, main ions, i and fast ions, f. The main ions have charge 1 and the fast ions charge Z. Quasineutrality requires:

$$n_e = n_i + Zn_f \tag{8.1}$$

Let us now introduce the fast fraction ε_f such that

$$n_f = \varepsilon_f n_e \tag{8.2}$$

$$n_i = (1 - Z\varepsilon_f)n_e \tag{8.3}$$

We require (8.1) to hold also for perturbations. Thus

$$\delta n_e = \delta n_i + Z\delta n_f \tag{8.4}$$

we obtain after dividing by n_e and using (8.2) and (8.3)

$$\frac{\delta n_e}{n_e} = (1 - \varepsilon_f Z)\frac{\delta n_i}{n_i} + \varepsilon_f Z\frac{\delta n_f}{n_f} \tag{8.5}$$

Equation 8.5 is the basic charge balance equation which needs to be fulfilled regardless of which physics desceiption we use for electrons, ions and hot ions. If we now study the influence of fast ions on a mode associated with the bulk plasma we have $\omega_{Df} \gg \omega$. Then, using (5.30) or (6.140) for the fast ions we obtain:

$$\frac{\delta n_f}{n_f} = -Z\frac{e\phi}{T_z} \tag{8.6}$$

The Boltzmann response for fast ions means that they do not take part in the destabilizing process but just responds in an isothermal way to the potential perturbations. The instability growth rateb is reduced due to the factor $1 - \varepsilon_f Z$ in front of the main ion response. This is the dilution effect. The same principle, of course, applies if we use a fast particle response which includes the parallel resonance. If we now consider modes with $\omega \sim \omega_{Df}$, the fast particles can be destabilizing. We will consider such cases in the following.

8.4 Fishbone Type Modes

Fishbone type modes are basically new branches of MHD type modes, introduced by fast trapped particles. The pure MHD modes are fairly close to marginal stability in the sense that the destabilizing effects balance the Alfvén line bending effect to give a mode with eigenfrequency close to zero. For the fishbone type modes we are only interested in the trapped population of fast particles. For these we start from the gyrokinetic equation (5.28) and average it over the bounce motion as described in the section of trapped particle modes in Chap. 6. This means that the $v_{\|}$ parts vanish from (5.28) and the magnetic drift is replaced by the bounce averaged precession frequency. The fast particle response is then given by (5.30) where the magnetic drift in the denominator is replaced by its bounce average. Now treating the fast particles as an additional particle species in the quasineutrality condition we may derive a dispersion relation by the trial function method described in (6.69)–(6.70). Ignoring the Ω^2 part, the dispersion relation (6.72) is generalized to the form

$$i\Omega = \delta W_{MHD} + \delta W_f \tag{8.7}$$

Here δW_{MHD} is given by (6.73) for the MHD ballooning mode and by the left hand side of (6.95) for the internal kink mode. It was evaluated for the first time with toroidal effects in [37]. The kinetic part δW_f is due to the fast particle response as obtained from (5.28). For a slowing down distribution δW_f takes the form:

$$\delta W_f = \sqrt{2}\beta_f \frac{\alpha}{\varepsilon_{nf}}\left(1 - \frac{3}{2}\varepsilon_{nh}\right)\ln\left(\frac{\alpha - 1}{\alpha}\right) \tag{8.8}$$

where β_f is the beta of the fast particles, $\alpha = \omega/\omega_{Df}$, ω_{Df} is the precession frequency and $\varepsilon_{nf} = \omega_{Df}/\omega_{*f}$ where ω_{*f} is the diamagnetic drift frequency of the fast particles. For a Maxwellian distribution we have [13]:

$$\delta W_f = \frac{1}{2}\beta_f \frac{\alpha}{\varepsilon_{nf}}\left\{\frac{2n_f\left[1 - W\left(\sqrt{2\alpha}\right)\right] - \alpha\varepsilon_{nf} + \left[\alpha\varepsilon_{nf} - 1 - n_f(2\alpha - 1)\right]}{\left[1 - W\left(\sqrt{2\alpha}\right)\right]^2}\right\} \tag{8.9}$$

The difference between the distributions (8.8) and (8.9) are usually not very large. Both the real part of the eigenfrequency and the growthrate are of the order of the precession frequency. This is also true when we use the advanced reactive fluid model in Chap. 6 for the fast particles [27]. The threshold for the reactive fluid model does, however, differ significantly. For the distributions (8.8) and (8.9) we obtain the threshold by balancing the imaginary part of δW_f with $i\Omega$. For the slowing down distribution the threshold is:

$$\beta_{crit} = \frac{<\omega_{Df}>}{\omega_A}\frac{2s}{\pi^2 q^2 I_0} \tag{8.10}$$

Where I_0 proportional to R/r includes the bounce averaging of the driving pressure term and <> denotes bounce averaging of the magnetic drift frequency. The growth rate is

$$\gamma = <\omega_{Df}>\frac{\pi^2}{4}\frac{\beta - \beta_{crit}}{\beta_{crit}} \tag{8.11}$$

When the plasma is heated β_f increases until it exceeds β_{crit}. Then the fishbone mode goes unstable and rapidly reduces β_f to below the threshold. The heating then again increases β_f and the process repeats itself. This leads to a fishbone like oscillation.

8.5 Toroidal Alfvén Eigenmodes

We will now consider the potentially most dangerous type of eigenmode which is the discrete one. This type of mode is, in the simpliest description, not subject to continuum damping. As mentioned above the Toroidal Alfvén Eigenmodes

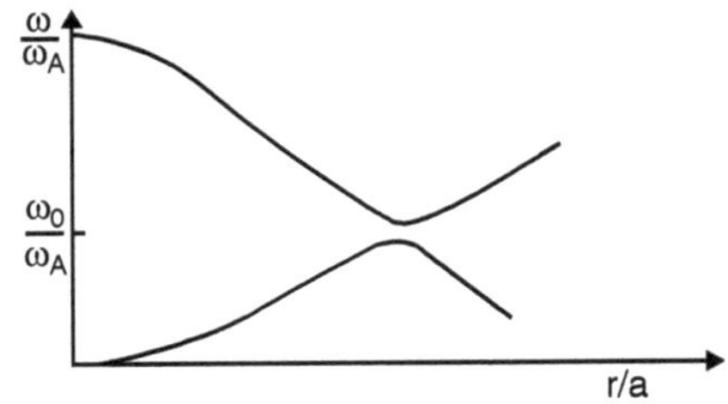

Fig. 8.1 Gap due to toroidal coupling of Alfvén waves with poloidal modenumbers m and m + 1

(TAE modes) occur in a gap in the Alfvén continuum, caused by a coupling of modes with poloidal mode number m and m + 1. In the cylindrical approximation $k_{||}$ is given by (6.7) i.e.

$$k_{||} = (m - nq)/qR \tag{8.12}$$

$$k_{||m} = -k_{||m+1} \tag{8.13}$$

Which happens when

$$q(r) = (2m + 1)/2n \tag{8.14}$$

In particular for m = −2 and n = −1 we obtain a resonance at q = 1,5. For this case we show the radial continuous spectra in Fig. 8.1. In the gap between the full lines there is a solution with only one (discrete) ω, which is the TAE mode. It has a real eigenfrequency close to the common cylindrical eigenfrequency ω_0 of the coupled Alfvén modes. Since this frequency is large, a kinetic resonance requires fast particles. The mathematical formulation usually makes use of the condition div**j** = 0. A convenient and rather general formulation was given in [23] as:

$$B\cdot\nabla\frac{1}{\mathrm{B}^2}B\cdot\nabla\Delta\phi + \frac{\omega(\omega-\omega_{*iT})}{\mathrm{v_A}^2}\Delta\phi - \frac{8\pi}{B^2}\frac{dP}{dr}k_\theta(\kappa\times\widehat{\mathbf{e}}_{||})\cdot\nabla\phi + \frac{4\pi}{B^2}\omega(\kappa\times B)\cdot\nabla(\delta\mathrm{P}_{||\mathrm{f}} + \delta\mathrm{P}_{\perp\mathrm{f}}) = 0 \tag{8.15}$$

Here $\delta P_{||f}$ and $\delta\, P_{\perp f}$ are the parallel and perpendicular components of the hot particle pressure tensor and the fact that only the curvature part of the magnetic drift appears is consistent with (6.21).

In (8.15) the first term is the shear Alfvén line bending term, the second term comes from the divergence of the polarisation and stress tensor drifts, the third term is the interchange term due to the driving bulk pressure and the fourth term is the driving term due to the pressure of the fast particles. An analytical expression for the growthrate of TAE modes, driven by only the parallel alpha particle pressure was obtained in [24].

$$\gamma = \omega_0 \frac{9}{4}\left[\beta_\alpha\left(\frac{\omega_\alpha}{\omega_0} - 0.5\right)F - \beta_e \frac{v_A}{v_e}\right] \tag{8.16}$$

Here β_α and β_e are the β values of alpha particles and electrons, respectively, and $\omega_{\bullet\omega}$ is the alpha particle diamagnetic drift frequency. $F(x) = x(1 + 2x^2 + 2x^4) \exp(-x^2)$, where $x = v_A/v_\alpha$represents the kinetic distribution. As pointed out above, we need particles with velocity close to the Alfvén velocity for a significant growthrate.

8.6 Discussion

We have here studied some of the most important collective effects associated with fast particles. These can be divided into on the one hand continuum modes and global modes and on the other hand modes driven by the perpendicular fast particle pressure through the resonance with the precession frequency of the trapped particles or by the parallel pressure through the Landau resonance with the circulating fast ions. As it turns out, the TAE mode, which normally is excited by the transit resonance, can also be destabilized by the precessing trapped ions [26].

Equation 8.15 can be used to describe both fishbone type ballooning modes and TAE modes (the only missing part is the kink term). The presence of the ion diamagnetic drift in the second term means that we can describe also the kinetic ballooning mode [23]. For perpendicular neutral beam heating, $\delta P_{\perp f}$ will dominate and a majority of the fast ions will be trapped. For parallel neutral beam injection $\delta P_{\parallel f}$ will dominate and the majority of the fast particles will be circulating. In the latter case we only have the TAE mode in the system. In the first case we can have both fishbone type modes and TAE modes. A major difference between these is that the fishbone mode is triggered close to marginal stability, i.e. only close to the MHD beta limit for the balloning type. On the other hand it can be excited for general fast ion precession frequency while the TAE mode requires particles with velocity close to the Alfvén velocity.

An investigation of thresholds for fishbone type and global modes in kinetic and reactive systems i.e. alternatively using (5.29) and (6.140) for the fast particles was made in [27].

Recent more detailed kinetic calculations of the stability of TAE's for reactors [28, 33, 35, 36] indicate that the modes may be somewhat more unstable than in TFTR although the situation should be possible to control.

Finally the nonlinear saturation of the instabilities is, of course, fundamental for the transport they cause. A usual estimate of the saturation level is obtained by balancing the linear growthrate and the $E \times B$ trapping frequency or nonlinear frequency shift. This leads to an estimate similar to (3.65). Berk and Breizman have investigated the details of the kinetic saturation process, including relaxation oscillations and flattening of the distribution function Refs. [29–31, 33, 34]

References

1. K.M. McGuire, R. Goldston, M.Bell et al., Phys. Rev. Lett. **50**, 891 (1983).
2. K.L. Wong, R.J. Fonck, S.F. Paul et.al., Phys. Rev. Lett. **66**, 1874 (1991).
3. K.L. Wong, R. Dust, R.J. Fonck et.al., Phys. Fluids **B4**,2122 (1992).
4. K.M. McGuire and the TFTR team, Phys Plasmas **2**, 2176 (1995)
5. D.S. Darrow, S.J. Zweben, S. Batha et. al. Phys. Plasmas **3** 1875 (1996).
6. H.L. Berk, Phys. Fluids **19**, 1255 (1976).
7. B. Coppi and F. Pegoraro, Comments Plasma Phys. Controlled Nuclear Fusion **5**, 131 (1979).
8. J. Weiland, Physica Scripta **23**, 801 (1981).
9. M.N. Rosenbluth and P.H. Rutherford, Phys. Rev. Lett. **34**, 1428 (1975).
10. K.T. Tsang, D.J. Sigmar and J.C. Whitson, Phys. Fluids **24**, 1508 (1981).
11. M.N. Rosenbluth, S.T. Tsai, J.W. Van Dam and M.G. Engquist, Phys. Rev. Lett. {\bf 51}, 1967 (1983).
12. L. Chen, R.B. White and M.N. Rosenbluth, Phys. Rev. Lett **51**, 1967 (1983).
13. R.B. White, L. Chen, F. Romanelli and R. Hay, Phys. Fluids **28**, 278 (1995).
14. J. Weiland and Liu Chen, Phys. Fluids **28**, 1359 (1986).
15. K. Appert, R. Gruber, F. Troyon and J. Vaclavic, J. Plasma Physics **24**, 1147 (1982).
16. D.W. Ross, G.L. Chen and S.M. Mahajan, Phys. Fluids **25**, 652 (1982).
17. C.Z. Cheng, L. Chen and M.S. Chance, Ann. Phys. **161**, 21 (1985).
18. C.Z. Cheng and M.S. Chance, Phys. Fluids **29**, 3695 (1986).
19. H. Biglari and L. Chen, Phys. Fluids **29**, 1760 (1986).
20. B. Coppi and F. Porcelli, Phys. Rev. Lett. **57**, 2277 (1986).
21. Y.M. Li, S.M. Mahajan and D.W. Ross, Phys. Fluids **30**, 1466 (1987).
22. J. Weiland, M. Lisak and H. Wilhelmsson, Physica Scripta **T16**, 53 (1987).
23. H. Biglari and L. Chen, Phys. Rev. Lett. **67**, 3681 (1991).
24. G.Y. Fu and J.W. Van Dam, Phys. Fluids **B1**, 1949 (1989).
25. G.Y. Fu and C.Z. Cheng, Phys. Fluids **B2**, 1427 (1990).
26. H. Biglari, F. Zonca and L. Chen, Phys. Fluids **B4**, 2385 (1992).
27. M. Liljeström and J. Weiland, Phys. Fluids **B4**, 630 (1992).
28. J. Candy and M.N. Rosenbluth, Nuclear Fusion **35**, 1069 (1995).
29. H.L. Berk and B.N. Breizman, Phys. Fluids **B2**, 2235 (1990).
30. H.L. Berk and B.N. Breizman, Phys. Fluids **B2** ,2246 (1990).
31. H.L. Berk, B.N. Breizman, J. Fitzpatrick, M.S. Pekker, H.V. Vong and K.L. Wong, Phys. Plasmas **3**, 1827 (1996).
32. T. Fülöp, M. Lisak, Ya. Kolesnichenko and D. Anderson, Plasma Phys. Control. Fusion **38**, 811 (1996).
33. B.N. Breizman, J. Candy, F. Porcelli and H.L. Berk, Phys. Plasmas **5**, 2326 (1998).
34. B.N. Breizman, H.L. Berk, J. Candy et. al., 17th IAEA Fusion Energy Conference, Yokohama, Japan, 1998, paper IAEA-F1-CN-69/TH2/4, (IAEA, Vienna 1999).
35. A. Jaun, A. Fasoli and W.W. Heidbrink, Phys. Plasmas **5**, 2952 (1998).
36. A. Jaun, K. Appert, T. Hellsten et.al., Phys. Plasmas **5**, 3801 (1998).
37. M.N. Bussac, R. Pellat, D. Edery and J.L. Soule, Phys. Rev. Lett. **35**, 1638 (1975).

Chapter 9
Nonlinear Theory

9.1 The Ion Vortex Equation

We have up to now mainly studied linear and quasilinear phenomena (with the exception for Sect. 6.10.4 in Chap. 6). Although quasilinear equations, in combination with an estimate of the saturation level, can be used to derive transport coefficients, it is important to go beyond this description in order to understand its region of applicability [1–83]. In particular nonlinear cascade rules [18, 20, 25, 26, 29, 30, 55] are important for the interplay between sources and sinks in k-space and the resulting saturation level and correlation length. We will thus here consider some simple nonlinear systems for turbulence in magnetized plasmas. We will also make a kinetic derivation of the diffusion coefficient which involves the turbulent transport itself as a decorrelation mechanism [3–5, 7, 8]. As we have pointed out in Chap. 3, the parallel ion motion may often be ignored in drift and flute modes. This is possible if $\omega \gg k_{\parallel} c_s$. For this case it is possible to derive a simple but still rather general nonlinear equation for the ion vorticity $\mathbf{\Omega} = \text{rot}\ \mathbf{v_i}$. We start from the fluid equation of motion for ions

$$\frac{\partial \mathbf{v}_i}{\partial t} + (\mathbf{v}_i \cdot \nabla)\mathbf{v}_i = \frac{e}{m_i}(\mathbf{E} + \mathbf{v}_i \times \mathbf{B}) - \frac{1}{m_i n}\nabla P_i + \mathbf{g} \tag{9.1}$$

Taking the curl of this equation, introducing

$$\Omega_{ci} = \frac{e}{m_i}\mathbf{B}$$

using the Maxwell equation

$$\nabla \times \mathbf{E} = -\frac{\partial \mathbf{B}}{\partial t}$$

J. Weiland, *Stability and Transport in Magnetic Confinement Systems*,
Springer Series on Atomic, Optical, and Plasma Physics 71,
DOI 10.1007/978-1-4614-3743-7_9,

and the vector relation

$$(\mathbf{v}\cdot\nabla)\mathbf{v}=\frac{1}{2}\nabla \mathrm{v}^2-\mathbf{v}\times\nabla\times\mathbf{v}$$

we obtain

$$\frac{\partial\boldsymbol{\Omega}_i}{\partial t}-\nabla\times(\mathbf{v}_i\times\boldsymbol{\Omega}_\mathbf{i})=-\frac{\partial\boldsymbol{\Omega}_\mathbf{ci}}{\partial t}+\nabla\times(\mathbf{v}_i\times\boldsymbol{\Omega}_\mathbf{ci})+\frac{1}{m_in^2}\nabla n\times\nabla P \tag{9.2}$$

When grad n $\times$ grad P $=0$ we may write (9.2) as

$$\frac{\partial}{\partial t}(\boldsymbol{\Omega}_i+\boldsymbol{\Omega}_\mathbf{ci})=\nabla\times[\mathbf{v}_i\times(\boldsymbol{\Omega}_\mathbf{i}+\boldsymbol{\Omega}_\mathbf{ci})] \tag{9.3}$$

If (9.3) is integrated around a closed line and we make use of Stokes theorem, we now obtain a generalized form of the familiar theorem of attachment of magnetic field lines to the plasma which reduces to the usual form when $\boldsymbol{\Omega}_\mathrm{i}\ll\boldsymbol{\Omega}_\mathrm{ci}$. In its usual form this theorem is primarily concerned with the perpendicular component of (9.3) while we will here be interested only in the parallel component of (9.3). Now since div$\boldsymbol{\Omega}_\mathbf{i}=$ div$\boldsymbol{\Omega}_\mathbf{ci}=0$ we find

$$\nabla\times(\mathbf{v}\times\boldsymbol{\Omega})=-\boldsymbol{\Omega}\nabla\cdot\mathbf{v}-(\mathbf{v}\cdot\nabla)\boldsymbol{\Omega}$$

where $\boldsymbol{\Omega}$ represents $\boldsymbol{\Omega}_\mathrm{i}$ or $\boldsymbol{\Omega}_\mathrm{ci}$. Now we realise that the operator $\boldsymbol{\Omega}\cdot\nabla$ represents a variation in the direction of the vorticity or the magnetic field. Since the vorticity will be due mainly to the $\mathbf{E}\times\mathbf{B}$ drift caused by the background magnetic field and we assume the magnetic perturbation to be small, we find that this operator represents a variation along the background magnetic field. Since the ion motion along the magnetic field is going to be neglected, we then drop this term. (The neglection of ion motion along **B** amounts to dropping $\mathrm{k}_\|\ \mathrm{c}_\mathrm{s}$, i.e. $\mathrm{k}_\|=0$). Then collecting terms and introducing

$$\frac{d}{dt}=\frac{\partial}{\partial t}+(\mathbf{v}\cdot\nabla)$$

we have

$$\frac{d}{dt}(\boldsymbol{\Omega}_i+\boldsymbol{\Omega}_\mathbf{ci})+(\boldsymbol{\Omega}_\mathbf{i}+\boldsymbol{\Omega}_\mathbf{ci})\nabla\cdot\mathbf{v}_\mathbf{i}=\frac{1}{m_in^2}\nabla n\times\nabla P \tag{9.4}$$

In order to express div $\mathbf{v}_\mathbf{i}$ we now use the ion continuity equation which may be written

$$\frac{dn_i}{dt} + n_i \nabla \cdot \mathbf{v_i} = 0$$

or

$$\nabla \cdot \mathbf{v_i} = -\frac{d}{dt} \ln n_i$$

then

$$\frac{d}{dt}(\boldsymbol{\Omega}_i + \boldsymbol{\Omega}_{\mathbf{ci}}) - (\boldsymbol{\Omega}_{\mathbf{i}} + \boldsymbol{\Omega}_{\mathbf{ci}})\frac{d}{dt} \ln n_i = \frac{1}{m_i n^2} \nabla n \times \nabla P \tag{9.5}$$

Since $\mathrm{div}\mathbf{v}_{\mathrm{E}} = 0$ it is often a good approximation to drop (d/dt) ln n_i completely. This corresponds to incompressible flow. The equation then describes the generation of vorticity $\boldsymbol{\Omega}_i$ and the magnetic field $\boldsymbol{\Omega}_i$ by the vector grad n $\times$ grad P. This vector is called the baroclinic vector and is present whenever there is an angle between the temperature and density gradient. It is one of the mechanism responsible for the generation of magnetic fields in laser pellet experiments. For the study of the ion vortex motion it is convenient to rewrite (9.5) in scalar form. We then note that the nonlinear E $\times$ B drift is given by (1.5), i.e. for $v_{\|} = 0$ the nonlinear contribution disappears and rot $\mathbf{v} = \Omega_{\|}$. We thus take the parallel component of (9.5). This means that we disregard perpendicular perturbations of $\boldsymbol{\Omega}_c$ in the ion equation. The approximation $k_{\|} = 0$ was also made in obtaining (4.31) from (4.20). We may rewrite (9.5) as:

$$\frac{d}{dt} \ln\left(\frac{\Omega_i + \Omega_{\mathbf{ci}}}{n_i}\right) = \frac{1}{m_i n^2} \frac{(\nabla n \times \nabla P) \cdot \hat{z}}{\Omega_i + \Omega_{ci}} \tag{9.6}$$

In order to treat consistently the ion temperature effects we have to include the velocity $\mathbf{v}_\pi$ due to the stress tensor. A correct evaluation of (9.6) in the presence of ion temperature gradient is then rather complicated. We will thus for simplicity assume the ion temperature to be small and drop the baroclinic vector. we then have the usually used form of the ion vortex equation

$$\frac{d}{dt} \ln\left(\frac{\Omega_i + \Omega_{\mathbf{ci}}}{n_i}\right) = 0 \tag{9.7}$$

We now write $\delta n_i = n_0 + \delta n_i$ where $\delta n_i \ll n_0$ and introduce the weak nonlinearity assumption $\Omega_i \ll \Omega_{ci}$. Then (9.7) takes the form

$$\frac{d}{dt}\left(\ln \Omega_{ci} + \frac{\Omega_i}{\Omega_{\mathbf{ci}}} - \ln\ n_0 - \frac{\delta n_i}{n_0}\right) = 0$$

Since we are now going to consider the ion temperature to be small we now use $\mathbf{v}_\mathrm{i} = \mathbf{v}_\mathrm{E} + \mathbf{v}_\mathrm{g}$ in (9.7). This is correct to first order in $\omega/\Omega_\mathrm{ci}$ since the ion inertia (polarisation drift) is already included in (9.7). We then find

$$\Omega_i = (\nabla \times \mathbf{v_E}) \cdot \widehat{\mathbf{z}} = \frac{1}{B_0}\Delta\phi \tag{9.8}$$

We can write (9.7) in the form

$$\begin{aligned}\left(\frac{\partial}{\partial t} + \mathrm{v_g}\frac{\partial}{\partial y}\right)\left(\frac{1}{B_0\mathbf{\Omega_{ci}}}\Delta\phi - \frac{\delta n_i}{n_0}\right) - \frac{1}{B_0}(\widehat{\mathbf{z}} \times \nabla\phi)\cdot\widehat{\mathbf{x}}\frac{\mathrm{d}}{\mathrm{dx}}\ln n_0 \\ = -\frac{1}{B_0}(\widehat{\mathbf{z}} \times \nabla\phi)\cdot\nabla\left(\frac{1}{B_0\mathbf{\Omega_{ci}}}\Delta\phi - \frac{\delta n_i}{n_0}\right)\end{aligned} \tag{9.9}$$

where we dropped both time and space derivatives of Ω_ci The grad B drift due to a variation of $\mathrm{B_0}$ along x may, however, be included in $\mathrm{v_g}$. We have now obtained a nonlinear equation for the ion dynamics. The density perturbation $\delta\mathrm{n_i}$ can be expressed in terms of φ by involving the electron dynamics and assumption of quasineutrality. We have in (9.9) dropped parallel ion motion, i.e. assumed $\mathrm{k}_\| \, \mathrm{c_s} \ll \omega$ which means that (9.9) is equivalent to the assumption $\mathrm{k}_\| = 0$ for the ions. For the electrons, however, we are still free to choose the region of interest. Remaining in the drift wave interval (3.5) we can use the Boltzmann distribution (3.2a) for the electrons. In combination with the quasineutrality condition this gives

$$\ln n_i = \ln n_0 + \frac{e\phi}{T_e}$$

which may be substituted directly into (9.7) without expansion. This means that for a Boltzmann distribution of electrons the electrons will not contribute to the nonlinearity. Then using the expanded form of $\ln(\mathbf{\Omega_i} + \Omega_\mathrm{ci})$ we obtain

$$\begin{aligned}\left(\frac{\partial}{\partial t} + \mathrm{v_g}\frac{\partial}{\partial y}\right)\left(\frac{1}{B_0\mathbf{\Omega_{ci}}}\Delta\phi - \frac{\delta n_i}{n_0}\right) - \frac{\kappa}{B_0}\frac{\partial\phi}{\partial y} \\ = -\frac{1}{B_0}(\widehat{\mathbf{z}} \times \nabla\phi)\cdot\nabla\left(\frac{1}{B_0\mathbf{\Omega_{ci}}}\Delta\phi - \frac{\delta n_i}{n_0}\right)\end{aligned} \tag{9.10}$$

We notice that the gravitational drift in (9.9) only gives a Doppler shifted frequency. Thus moving to the frame with velocity $\mathrm{v_g}$ we obtain the equation

$$\frac{d}{dt}\left(\frac{1}{B_0\mathbf{\Omega_{ci}}}\Delta\phi - \frac{\delta n_i}{n_0}\right) - \frac{\kappa}{B_0}\frac{\partial\phi}{\partial y} = -\frac{1}{{B_0}^2\mathbf{\Omega_{ci}}}(\widehat{\mathbf{z}} \times \nabla\phi)\cdot\nabla\Delta\phi + \frac{1}{B_0 T}\frac{dT_e}{dx}\frac{e\phi}{T_e}\frac{\partial\phi}{\partial y}$$

where we included the possibility of an electron temperature gradient in the x direction. We may compare the two nonlinear terms in the following way. Introducing $T/eB_0 = c_s^2/\Omega_{ci}$, we can rewrite the first term as $1/B_0\rho^2\left(\text{grad}(e\varphi/T_e)x\hat{\mathbf{z}}\right) \bullet \text{grad}\Delta\varphi$. We then arrive at the ratio $(d/dx)\ln T_e/k_y k^2\rho^2$ between the nonlinear terms. It is natural to assume $(d/dx)\ln T \ll k_y$. Since, however, $k^2\rho^2$ may be small, the second nonlinear term will not always be negligible. We will, nevertheless, in the following assume this to be the case. We then arrive at the *Hasegawa-Mima* equation (the quasi-geostrophic vortex equation)

$$\frac{d}{dt}(\rho^2\Delta\phi - \phi) - v_{*e}\frac{\partial\phi}{\partial y} = \frac{\rho_s{}^2}{B_0}(\nabla\phi \times \hat{\mathbf{z}}) \cdot \nabla\Delta\phi \tag{9.11}$$

Where we multiplied by T_e/e. Equation 9.11 is the small gyroradius limit of (5.43) which was derived from the nonlinear gyrokinetic equation. We again observe that the interchange frequency is absent when the electrons obey a Boltzmann distribution. Equation 9.11 has the conserved quantities

$$W = \int \left[\rho_s{}^2\Delta\phi^2 + \phi^2\right] d^3r \tag{9.12}$$

$$V = \int \left[\rho_s{}^2\Delta\phi^2 + \left(\rho^2\Delta\phi^2\right)^2\right] d^3r \tag{9.13}$$

Where W is the energy and V is the enstrophy (squared vorticity). We notice that in the linear approximation, (9.11) reduces to

$$\omega = \frac{\omega_{*e}}{1 + k_y{}^2\rho_s{}^2}$$

Which is the usual dispersion relation for electron driftwaves without parallel ion motion.

We now turn to the case $\omega \gg k_{\|}v_{the}$. This is the limit of electrostatic interchange modes. In this limit the electrons are not Boltzmann distributed. Instead, we may use the approximation $k_{\|} = 0$ also for electrons. The electrons can then be described by the continuity equation

$$\frac{\partial n_e}{\partial t} + \frac{\kappa}{B_0}\frac{\partial\phi}{\partial y} = \frac{1}{B_0}(\nabla\phi \times \hat{\mathbf{z}}) \cdot \nabla n_e \tag{9.14}$$

where we used $\mathbf{v}_e = \mathbf{v}_E$ since temperature effects do not enter into the continuity equation. In the limit $\kappa \ll k$ we can write $\text{grad}\, n_e/n_0 = \text{grad}\, \delta n_e/n_0$.

Thus dividing (9.14) by n_0 and subtracting it from (9.9) we arrive at the coupled system of equations

$$\frac{1}{B_0\Omega_{ci}}\left(\frac{\partial}{\partial t}+v_g\frac{\partial}{\partial y}\right)\Delta\phi - v_g\frac{\partial}{\partial y}\frac{\delta n_i}{n_0} = \frac{1}{B_0{}^2\Omega_{ci}}(\nabla\phi\times\hat{\mathbf{z}})\cdot\nabla\Delta\phi \tag{9.15}$$

$$\frac{d}{dt}\left(\frac{\delta n}{n_0}\right)+\frac{\kappa}{B_0}\frac{\partial\phi}{\partial y} = \frac{1}{B_0}(\nabla\phi\times\hat{\mathbf{z}})\cdot\nabla\frac{\delta n}{n_0} \tag{9.16}$$

where we again used the quasineutrality condition and introduced $\delta n \equiv \delta n_e = \delta n_i$. We notice that (9.15) couples to (9.16) only due to the gravitational drift v_g. In the linear approximation we obtain from (9.16)

$$\frac{\delta n}{n_0} = \frac{\kappa}{B_0\omega}k_y\phi$$

which is the same relation as (3.22). Substituting this expression into (9.15) we find the dispersion relation

$$\omega(\omega - k_y v_g) + \kappa g k_y{}^2/k^2 = 0 \tag{9.17}$$

where we introduced $v_g = -g/\Omega_{ci}$ and $k^2 = k_x{}^2 + k_y{}^2$. This dispersion relation is identical to (3.23) if $g \rightarrow g_i + (m_e/m_i)g_e$. It is also of interest to note that when $g = v_{th}{}^2/R_c$, i.e. is due to the curvature, the gravitational drift of the electrons is of the same order as that of the ion for equal temperatures. For this case we should replace g by $g_i + (m_e/m_i)g_e$ which will, however, only modify the interchange frequency $(\kappa g)^{1/2}$ by a factor $\sqrt{2}$.

The simplest nonlinear process described by (9.11) is the three wave interaction. We write

$$\phi = \sum \phi_k(t)e^{i(k_x x + k_y y - \omega t)} + c.c \tag{9.18}$$

Substituting (9.18) into (9.11) we now obtain

$$\begin{aligned}&\left(-i\omega+\frac{d}{dt}\right)(1+k^2\rho^2)\phi_k - ik_y v_*\phi_k\\ &=\frac{\rho_s{}^2}{B_0}\left[(\mathbf{k_1}\times\hat{\mathbf{z}})\cdot\mathbf{k_2}k_2{}^2+(\mathbf{k}_2\times\hat{\mathbf{z}})\cdot\mathbf{k}_1k_1{}^2\right]\phi_{k_1}\phi_{k_2}e^{i(\omega-\omega_1-\omega_2)t}\end{aligned} \tag{9.19}$$

where we assumed the matching condition

$$\mathbf{k} = \mathbf{k}_1 + \mathbf{k}_2$$

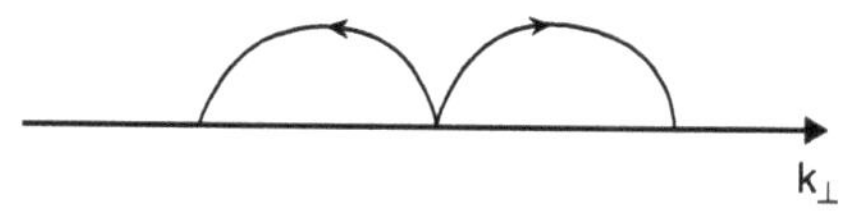

Fig. 9.1 Dual cascade

to be fulfilled. Assuming now that ω is a solution of the linear dispersion relation, we obtain the three coupled equations

$$\frac{\partial \phi_k}{\partial t} = \frac{{\rho_s}^2}{B_0(1+k^2\rho^2)}(\mathbf{k_1} \times \mathbf{k_2}) \cdot \widehat{\mathbf{z}}({k_2}^2 - {k_1}^2)\phi_{k_1}\phi_{k_2}e^{i\Delta\omega t} \quad (9.20)$$

$$\frac{\partial \phi_{k_1}}{\partial t} = -\frac{{\rho_s}^2}{B_0(1+{k_1}^2\rho^2)}(\mathbf{k} \times \mathbf{k_2}) \cdot \widehat{\mathbf{z}}({k_2}^2 - k^2)\phi_k{\phi_{k_2}}^* e^{-i\Delta\omega t} \quad (9.21)$$

$$\frac{\partial \phi_{k_2}}{\partial t} = -\frac{{\rho_s}^2}{B_0(1+{k_2}^2\rho^2)}(\mathbf{k} \times \mathbf{k}_1) \cdot \widehat{\mathbf{z}}({k_1}^2 - k^2)\phi_k{\phi_{k1}}^* e^{-i\Delta\omega t} \quad (9.22)$$

Where we introduced the frequency mismatch $\Delta\omega = \omega - \omega_1 - \omega_2$.

We can now substitute the matching condition for the wave vectors into (9.22) in order to eliminate **k**. The vector products may then all be expressed in terms of $(\mathbf{k}_1 \times \mathbf{k}_2) \cdot \dot{\mathbf{z}}$. This leads, however, to a change of sign in all three equations. This means that the coupling factor of (9.21) will have opposite sign to the coupling factors of the other two equations. In this situation mode 1, i.e. the mode with the immediate magnitude of k will act as a pump wave and we have cascading of wave quanta towards smaller and larger k according to Fig. 9.1.

The threshold for parametric interaction is

$$|\phi_{k_1}|^2 > \frac{\Delta\omega^2}{4V_0V_2} \quad (9.23)$$

Where V_0 and V_2 are the coupling factors. By using the dispersion relation in the form $k^2\rho^2 = k_y v_{\bullet e}/\omega_k - 1$, it is possible to write

$$\begin{aligned} {k_1}^2 - {k_2}^2 &= -\omega M \\ k^2 - {k_1}^2 &= \omega_2 M \\ {k_2}^2 - k^2 &= \omega_1 M \end{aligned} \quad (9.24)$$

where

$$M = \frac{v_{*e}}{3\rho^2\omega\omega_1\omega_2}\left[\omega(k_{y2} - k_{y1}) + \omega_1(k_y + k_{y2}) - \omega_2(k_y + k_{y1})\right]$$

We then find that when $k^2\rho^2 \ll 1$, i.e. $\Delta\omega \to 0$, the pump wave will be the mode with the largest frequency.

Assuming the presence of a large amplitude long wavelength mode Hasegawa and Mima derived, for a random phase situation, a stationary spectrum of the form

$$\left|\frac{e\phi_k}{T_e}\right|^2 = \Gamma \frac{(k\rho)^{\alpha}}{(1+k^2\rho^2)^{\beta}} \tag{9.25}$$

where $\alpha = 1.8$ and $\beta = 2.2$. This spectrum is in reasonable agreement with spectra observed in tokamak experiments where $10^6 < \Gamma < 10^{-5}$. Computer investigations by Fyfe and Montgomery [14] show a spectrum with the dependence $k^{-14/3}$ below and k^{-6} above a source while recent experiments give the variation $k^{-3.5}$. The experiments, however, include several effects not included in (9.25) such as ion temperature effects and linear damping or growth. Another important phenomenon observed in nonlinear simulations of (9.11) is the generation of zonal flows [19]. Such flows may cause a stabilization of drift wave turbulence, leading to internal transport barriers.

For the system (9.15) and (9.16) the derivation of the coupling factors is considerably more complicated. The result may be written in the form

$$\frac{\partial \phi_k}{\partial t} = V_{12}\phi_{k_1}\phi_{k_2}e^{i\Delta\omega t} \tag{9.26}$$

where

$$V_{12} = \rho_s{}^2 \frac{(\mathbf{k_1}\times\mathbf{k_2})\cdot\widehat{\mathbf{z}}}{k^2} \times \cdot\widehat{\mathbf{z}}\left[k_2{}^2 - k_1{}^2 - \rho^4 k^2 k_1{}^2 k_2{}^2 \frac{\omega\omega_1\omega_2}{\omega'\omega_1{}'\omega_2{}'}\left(\frac{\mathrm{v_g}}{\mathrm{v_{*e}}}\right)^2\left(\frac{\omega_2}{\omega_2{}'} - \frac{\omega_1}{\omega_1{}'}\right)\right]\frac{\omega-\omega'}{2\omega-\omega'} \tag{9.27}$$

Where $\omega' = k_y v_g$. The last factor gives the sign of the energy. The system described by (9.27) has the same cascade rules determined by k numbers as (9.20)–(9.22). This is, however, no transition to the usual weak turbulence rule for perfect matching. The wave energy is given by

$$W_k = k^2\rho_s{}^2\left(\frac{e\phi_k}{T_e}\right)^2 + \frac{\mathrm{v_g}}{\mathrm{v_{*e}}}\left(\frac{n_k}{n_0}\right)^2 = \frac{2\omega-\omega'}{\omega-\omega'}k^2\rho_s{}^2\left(\frac{e\phi_k}{T_e}\right)^2$$

This wave energy is contrary to the usual weak turbulence case conserved also in the presence of mismatch. We also note that this expression for the wave energy, obtained from a nonlinear conservation relation in a fluid model, agrees with the expression in (4.79) obtained from a linear kinetic theory, to first order in the FLR parameter.

From the ion vortex equation we may derive a simple condition for the applicability of quasi-neutrality. Using the Poisson equation in (9.7) we have

$$\frac{d}{dt}\ln\left(\frac{\Omega_i + \mathbf{\Omega_{ci}}}{n_e - \varepsilon_0 \Delta\phi/e}\right) = 0 \tag{9.28}$$

Since here n_e is the total electron density we can expand the denominator for $n_e \gg \varepsilon_0\Delta\varphi/e$. Using (9.28) for Ω_i and assuming $\Omega_{ci} \gg \Omega_i$ we obtain, dropping $\Omega_i\varepsilon_0\Delta\varphi/e$

$$\frac{d}{dt}\ln\left[\left(\frac{1}{B_0}\Delta\phi\left(1 + \frac{{\Omega_{ci}}^2}{{\omega_{pi}}^2}\right) + \mathbf{\Omega_{ci}}\right)/n_e\right] = 0 \tag{9.29}$$

We thus find the condition ${\omega_{pi}}^2 \gg {\Omega_{ci}}^2$ for quasineutrality. For a tokamak plasma we have typically $\omega_{pi} \sim 40\Omega_{ci}$ so the condition for quasineutrality is well fulfilled.

9.2 The Nonlinear Dielectric

An alternative to the previous formulation of the nonlinear dynamics in terms of the ion vortex equation is the formulation in terms of a dielectric function. For electrostatic modes this is

$$\omega\varepsilon(\omega, k)E_{\omega,k} = -\frac{i}{\varepsilon_0}{j_{\omega,k}}^{(2)} \tag{9.30}$$

where $\varepsilon(\omega,k)$ is the linear dielectric function given by (4.66). For $k^2\lambda_{De}^2 \ll 1$ and $T_i \ll T_e$ we obtain

$$\varepsilon(\omega, k) = \frac{1}{k^2\,{\lambda_{de}}^2}\left(1 + k^2\rho^2 - \frac{\omega_{*e}}{\omega}\right) \tag{9.31}$$

The current $j^{(2)}$ is the nonlinear current. For electrostatic drift waves

$${j_{\omega,k}}^{(2)} = en{\mathbf{v_{pi}}}^{(2)} = -\frac{en}{B\Omega_{ci}}(\widehat{\mathbf{z}} \times \nabla\phi)\cdot\nabla\nabla\phi \tag{9.32}$$

i.e. the nonlinear part of the polarisation drift neglecting $\mathbf{v}_{*i}$ when $T_i \ll T_e$. Substitution into (~9.30) leads t

$$\left[\omega(1 - \rho^2\Delta) - i\mathbf{v_{*e}}\cdot\nabla\right]\nabla\phi = -i\Delta\frac{T_e}{eB\Omega_{ci}}(\widehat{\mathbf{z}} \times \nabla\phi)\cdot\nabla\nabla\phi \tag{9.33}$$

Taking the divergence of (9.33) we obtain

$$\left[\omega(1-\rho^2\Delta) - i\mathbf{v}_{*\mathbf{e}}\cdot\nabla\right]\Delta\phi = -i\rho^2\Delta(\widehat{\mathbf{z}}\times\nabla\phi)\cdot\nabla\Delta\phi$$

Then inverting the Laplacian and transforming $\omega \rightarrow$ id/dt we obtain the Hasegawa-Mima equation

$$\frac{d}{dt}(\rho^2\Delta\phi - \phi) - \mathbf{v}_{*e}\cdot\nabla\phi = \frac{\rho_s{}^2}{B_0}(\nabla\phi\times\widehat{\mathbf{z}})\cdot\nabla\Delta\phi \tag{9.34}$$

Which is identical to (9.11).

We note that the particularly simple frequency dependence of (9.31) made it possible to transform to the time domain without expanding $\varepsilon(\omega,k)$ around a linear eigenfrequency. Because of this (9.34) is valid in the strongly nonlinear regime.

9.3 Diffusion

The main reason for the interest in collective perturbations in magnetized plasmas is the anomalous transport caused by a turbulence of such perturbations. The low frequency vortex modes treated here are of special interest for several reasons. First we observe that a convection across the magnetic field is associated with the vorticity. Second as we will see in this section low frequency modes cause efficient transport. Third, these modes are frequently driven unstable by inhomogeneities in pressure and magnetic field, making them hard to avoid in a confined plasma. Although the anomalous transport is of convective type it is usually treated as a diffusive process, This can be justified in a turbulent state where the particle motion in the wave fields is stochastic and the requirement on particle stochasticity is in fact more easily fulfilled than the random phase approximation for the waves. For a stochastic motion of particles the diffusion coefficient is usually defined as

$$D = \lim\frac{1}{2t}<\Delta r^2(t)>;..t\rightarrow\infty \tag{9.35}$$

where Δr is the distance from the point where the particle was at t = 0 and <> denotes an average over all possible initial velocities or more generally an ensemble average. We now introduce the velocity v(t) so that

$$\Delta r(t) = \int_0^t \mathrm{v}(\mathrm{t}')\mathrm{dt}'$$

and

$$D = \lim \frac{1}{2t} \left\langle \int_0^t dt' \int_0^t \mathrm{v(t')v(t'')} dt'' \right\rangle; ..t \to \infty$$
$$= \lim \frac{1}{2t} \int_0^t dt' \int_0^t <\mathrm{v(t')v(t'')}>\mathrm{dt''}; ..t \to \infty$$

We shall now assume that we have a stationary stochastic process such that $<\mathrm{v(t')v(t'')}> = <\mathrm{v(t' - t'')v(0)}>$. This means that the correlation between the velocities only depends on the difference in time $\tau = \mathrm{t'} - \mathrm{t''}$ and

$$D = \lim \frac{1}{2t} \int_0^t dt' \int_0^t <\mathrm{v}(\tau)\mathrm{v}(0)>\mathrm{dt''}; ..t \to \infty$$

which simplifies to

$$D = \int_0^\infty <\mathrm{v}(\tau)\mathrm{v}(0)>\mathrm{d}\tau \tag{9.36}$$

Another usual way of defining D is as the coefficient in the diffusion equation

$$\frac{\partial^2 n}{\partial t^2} = D \frac{\partial^2 n}{\partial r^2} \tag{9.37}$$

Where n = n(r,t) is the particle density. A solution to (9.37) corresponding to the initial state where all particles are collected at r = 0 is

$$\tilde{n}(r,t) = \frac{N}{(4\pi D t)^{1/2}} e^{-r^2/4Dt} \tag{9.38}$$

Where N is the total number of particles. Clearly the possibility of finding a particle between r and r + Δr at time t is ñ(r,t)Δr/N if Δr is small enough. This means that the ensemble average of a quantity Q(r,t) can be written

$$<Q> = \frac{1}{N} \int_{-\infty}^{\infty} \tilde{n}(r,t) Q(r,t) dr \tag{9.39}$$

As is easily seen we now have

$$<r^2> = \frac{1}{N} \int_{-\infty}^{\infty} \tilde{n}(r,t) r^2 dr = 2Dt$$

Since r here is the total deviation in position since t = 0 we realise that the two definitions of D are equivalent. In order to derive a useful expression for D for a time and space dependent process it is convenient to start from (9.36) where $v(\tau)$ is represented in Fourier form

$$v(\tau) = v(r(\tau),\tau) = \frac{1}{(2\pi)^3}\int v_{k,\omega} e^{i[\omega\tau - \mathbf{k}\cdot\mathbf{r}(\tau)]} d\omega d\mathbf{k}$$

where the two space dimensions were assumed.

We then obtain from (9.36)

$$D = \int_0^\infty d\tau \frac{1}{(2\pi)^3}\int <|v_{k,\omega}|^2> e^{i\omega\tau} <e^{-i\mathbf{k}\cdot\mathbf{r}(\tau)}> d\omega d\mathbf{k} \tag{9.40}$$

where we assumed that $v_{k,\omega}$ is uncorrelated with the phase function. In order to obtain D we now need to know the velocity spectrum and the ensemble average of the space phase function. The latter can be obtained by using the representation (7.39) of the ensemble average. This leads to the result

$$<e^{-i\mathbf{k}\cdot\mathbf{r}}> = e^{-k^2Dt} \tag{9.41}$$

This result was verified numerically for thermal equilibrium by Joyce, Montgomery and Emery [10]. The characteristic time $(k^2D)^{-1}$ is usually called the orbit decorrelation time and is the time after which an average particle has moved so far due to diffusion that the field is uncorrelated with the field at the initial point. Specializing now to resonant modes where $\mathbf{v}_{k\omega} = \mathbf{v}_k\delta(\omega - \omega(\mathbf{k}))$ where $\omega(\mathbf{k})$ is the solution of a dispersion relation we find

$$D = \int_0^\infty d\tau \frac{1}{(2\pi)^3}\int <v_{k,\omega}{}^2> e^{i\omega(k)\tau - k^2D\tau} d\omega d\mathbf{k} = \frac{1}{(2\pi)^2}\int \frac{<v_k{}^2>}{-i\omega(k) + k^2D} dk \tag{9.42}$$

where we assumed convergence at $\tau = \infty$, i.e. $\mathrm{Im}\,\omega < k^2D$. We note that (9.40) and (9.42) contain integration over a nonlinear particle orbit in the diffusive limit if the diffusion is due to turbulence. Introducing now

$$\omega(k) = \omega_{kr} + i\gamma_k$$

and the reality condition $\omega_{kr} = -\omega_{-kr}$ we obtain

$$D = \frac{1}{(2\pi)^2}\int <v_k{}^2> \frac{k^2D + \gamma_k}{\omega_{kr}{}^2 + (k^2D + \gamma_k)^2} dk \tag{9.43}$$

Equation 9.43 shows that the orbit decorrelation and the wave growth both contribute to the transport while the real part of the eigenfrequency decreases the transport. for low frequency modes the dominant convective velocity is the **E** × **B** drift velocity. We then have

$$\mathbf{v_k} = \frac{i}{B_0}(\widehat{\mathbf{z}} \times \vec{\mathbf{k}})\phi_{\mathrm{k}}$$

The most efficient mode in a plasma in a homogeneous magnetic field is the convective cell mode. For this mode ω_{kr} and the orbit decorrelation usually dominates the damping. In this case we can solve (9.43) for D with the result

$$D = \frac{1}{B_0}\left(\int \frac{1}{2\pi} |\phi_k|^2 \, d\mathbf{k}\right)^{1/2} \tag{9.44}$$

Which is the diffusion coefficient for convective cells. It was first derived by Taylor and Mc Namara [7]. For a thermal equilibrium spectrum in the two dimensional case

$$\frac{k^2|\phi_k|^2}{8\pi} = \frac{T}{2\varepsilon} \tag{9.45}$$

Where ε is the dielectric function. We thus obtain

$$D = \frac{1}{B_0}\left(\frac{2\mathrm{T}}{\varepsilon} \ln \frac{Lk_{\max}}{2\pi}\right)^{1/2} \tag{9.46}$$

where L is the maximum allowed wavelength (system dimension). The influence of ε was introduced by Okuda and Dawson (9.8). The dielectric constant used was (compare Eq.4.64).

$$\varepsilon = 1 + \frac{\omega_{pe}{}^2}{\Omega_{ce}{}^2} + \frac{\omega_{pe}{}^2}{\Omega_{ce}{}^2}$$

which leads to a Bohm like diffusion D ~ 1/B for $\omega_{\mathrm{pi}}{}^2/\omega_{\mathrm{ci}}{}^2 \ll 1$ and to a diffusion independent of B for $\omega_{\mathrm{pi}}{}^2/\omega_{\mathrm{ci}}{}^2 \gg 1$. This diffusion is, in the plateau regime, comparable to the classical diffusion but much larger in the Bohm regime. Most fusion machines are supposed to work in the plateau regime but also here the anomalous transport will dominate in a turbulent state where the excitation level will be much larger than that given by (9.45).

Another mode of considerable interest is the magnetostatic mode (see Sect. 5.1.2). This mode is electromagnetic and causes mainly electron diffusion by perturbing the magnetic flux surfaces. The velocity in (9.43) is here given by

$$\mathrm{v_k} = \mathrm{v}_{||} \frac{\delta \mathbf{B}_{\perp}}{\mathrm{B}_0}$$

where $\delta\mathbf{B}_\perp$ is the perturbation of the magnetic field perpendicular to the background magnetic field and $v_{||}$ is the thermal velocity. This process was studied by Chu, Chu and Ohkawa ([19]) where the diffusion coefficient

$$D = \frac{T}{B_0}\left[\frac{2}{mL_{||}}\ln\left(\frac{Lk_{\max}}{2\pi}\right)\right]^{1/2} \tag{9.47}$$

was obtained for a thermal equilibrium. Here $L_{||}$ is the system length parallel to the magnetic field and L is the dimension in the perpendicular direction. This diffusion coefficient has a Bohm like T/B scaling. Since this is mainly an electron diffusion, charge separation effects will efficiently prevent it from leading to actual particle transport. It will, however, instead cause a thermal conductivity and it has been suggested that processes of this kind could explain the anomalous thermal conductivity of tokamaks which is about two orders of magnitude larger than the classical. In the derivations of the diffusion coefficients (9.46) and (9.47) it was assumed that the real part of the eigenfrequency could be neglected. This is not always a realistic assumption. For the convective cell mode curvature of the magnetic field lines can violate this assumption while for the magnetostatic mode a density inhomogeneity is enough. For both modes magnetic shear can limit the maximum perpendicular extension of the mode. In such situations nonlinear modes driven by the ponderomotive force may sometimes be more dangerous.

9.4 Fokker-Planck Transition Probability

The use of the solution of the diffusion equation for calculating ensemble averages can be generalised to solutions of the Fokker-Planck equation for diffusion in phase [81]. We consider solutions of the equation:

$$\left(\frac{\partial}{\partial t} + \mathbf{v}\cdot\frac{\partial}{\partial \mathbf{r}}\right) W(X,X',\mathrm{t},\mathrm{t}') = \frac{\partial}{\partial \mathbf{v}}\left[\beta\mathbf{v} + D^{\mathrm{v}}\frac{\partial}{\partial \mathbf{v}}\right] W(X,X',\mathrm{t},\mathrm{t}') \tag{9.48}$$

Where $X = (\mathbf{r},\mathbf{v})$ is the phase space coordinate, the diffusion coefficient in velocity space D^{v} is, in general, a tensor and β is the friction coefficient. We can see (9.48) as the generalisation of (9.37) to include also velocity space, i.e. we now consider the six dimensional phase space. W is here called the transition probability and is used to calculate ensemble averages in a way analogous to (9.39). We also note that (9.48) was derived for turbulent collisions [81]. This means that the friction and diffusion coefficients have the general forms:

$$\beta = \sum_k \beta_k |\phi_k|^2$$

$$D^{v} = \sum_{k} d_k |\phi_k|^2$$

Equation 9.48 has solutions of the form (2.1)

$$W(X, X', \tau; \tau') = \frac{e^{3\beta t}}{8\pi^3 \Delta^{3/2}} \exp\left[-\frac{1}{2\Delta} (a_{ij}\delta\rho_i\delta\rho_j + 2h_{ij}\delta\rho_i\delta P_j + b_{ij}\,\delta P_i\,\delta P_j) \right] \quad (9.49)$$

Where

$$\Delta = \frac{1}{3}(a_{ij}b_{ij} - h_{ij}h_{ji})$$

$$a_{ij} \equiv a_{ij}(t, t') = \frac{2}{\beta^2} \int_{t'}^{t} D_{i,j}{}^{v}(s)ds$$

$$b_{ij} \equiv b_{ij}(t, t') = 2 \int_{t'}^{t} D_{ij}{}^{v}(s) e^{2\beta(s-\tau')} ds$$

$$h_{ij} \equiv h_{ij}(t, t') = -\frac{2}{\beta^2} \int_{t'}^{t} D_{ij}{}^{v}(s) e^{\beta(s-\tau')} ds$$

$$\delta\boldsymbol{\rho} = \mathbf{v}e^{\beta\tau} - \mathbf{v}' \quad \tau = t - t'$$

$$\delta\mathbf{P} = \mathbf{r} - \mathbf{r}' + \frac{\mathbf{v} - \mathbf{v}'}{\beta}$$

For the one dimensional case with time independent diffusion coefficient we obtain

$$a = \frac{2}{\beta^2} D^{v}\tau \quad b = \frac{1}{\beta} D^{v}(e^{2\beta\tau} - 1)$$

And

$$h = -\frac{2}{\beta^2} D^{v}(e^{\beta\tau} - 1) \quad \Delta = ab - h^2$$

$$W(X, X', \tau; \tau') = \frac{e^{\beta t}}{2\pi\Delta^{1/2}} e^{-\frac{1}{2\Delta}(a\delta\rho^2 + 2h\delta\rho\delta P + b\delta P^2)} \quad (9.50)$$

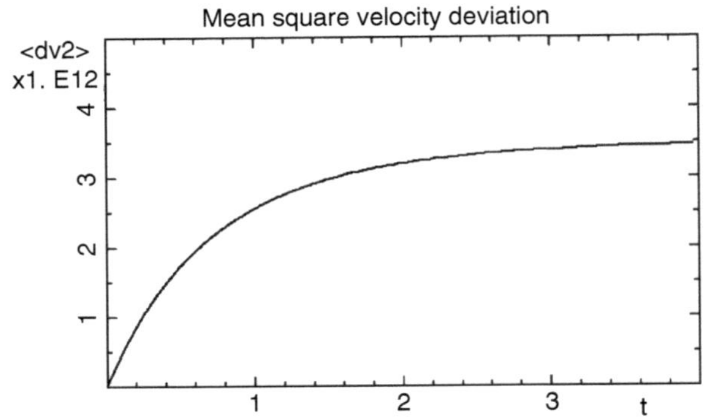

Fig. 9.2 Time variation of $<\Delta v^2>$ as given by the Fokker Planck equation

We may see (9.50) as a weight function to derive ensemble averages. Some examples are

$$<\Delta r> = \frac{v}{\beta}(1 - e^{-\beta\tau})$$

$$<e^{-i\mathbf{k}\cdot\Delta\mathbf{r}}> = \exp\left[i\frac{kv}{\beta}(1 - e^{-\beta\tau}) - \frac{k^2}{\beta^2}\int_0^\tau d\xi D^v(t-\xi)(1 - e^{-\beta(\tau-\xi)})^2\right]$$

In the stationary case we have

$$<e^{-i\mathbf{k}\Delta\cdot\mathbf{r}}> = \exp\left(ikv\tau - \frac{k^2 D^v}{3}\tau^3\right)(\beta\tau \ll 1) \tag{9.51}$$

$$<e^{-i\mathbf{k}\Delta\cdot\mathbf{r}}> = \exp\left(\frac{ikv}{\beta} - k^2 D\tau\right)(\beta\tau \gg 1) \tag{9.52}$$

Where $D = D^v/\beta$ is the diffusion coefficient in configuration space.

We here recognize the τ^3 dependence found by Dupree and Weinstock [4, 5] by renormalization in (9.51) and the diffusivity in ordinary space, i.e. (9.41) in (9.52).

However we furthermore get

$$<\Delta v^2> = \frac{D^v}{\beta}(1 - e^{-2\beta\tau}) + v_0{}^2(1 - e^{-\beta\tau})^2 \tag{9.53}$$

Where $\mathbf{v_0}$ is a fixed initial condition which we will choose to be zero. We then notice that (9.53) gives the usual diffusion in velocity space for small times while $<\Delta v^2>$ saturates for $\tau > 1/\beta$. The time dependence is given by Fig. 9.2

As pointed out in Chap. 6 the saturation occurs at $t \approx \beta^{-1}$. This is clear from (9.53). We also note that the friction enters as a complex nonlinear frequency shift which is expected to wipe out wave particle resonances, as discussed in Sect. 6.10.4. We may now obtain the corresponding solution in the non-Markovian case as a convolution in time of (9.50). It can be rewritten in terms of Fourier components in time of D^{v} (t,τ) and $\beta(t,\tau)$. From this formulation the diffusion coefficient (3.67) for diffusion in real space emerges in a natural way [81]. The result obtained in [81] is, however, more general since it includes the nonlinear frequency shift.

9.5 Discussion

In this chapter we have derived the general form of the ion vortex equation which can be used to describe most types of vortex modes in plasmas as well as in fluids. Here we used it to derive nonlinear equations for drift waves and interchange modes. For these types of modes we discussed the dual cascade towards shorter and longer space scales, typical of two dimensional systems.

The cascade towards longer space scales is particularly important for transport and we generally need some damping mechanism for long wavelengths to obtain a realistic level of the transport. This mechanism will most likely be sheared plasma flows generated nonlinearly or by neutral beams or neoclassical effects.

These flows may create an absorbing boundary condition for long wavelengths if sufficiently long wavelengths are included in the system, as discussed in Sect. 6.10.5. We also note the discussion of conservation relations and the comparison between the expressions for the wave energy of interchange modes obtained here and from the dielectric properties in Chap. 4.

The calculation of diffusion from particle orbit integrations is a complement to the quasilinear calculations in Chap. 3. We note the convenient use of the solution of the diffusion equation as a weight function (transition probability) for calculating ensemble averages. This method was later extended to the general Fokker-Planck equation for diffusion in velocity space. From this calculation the renormalization by Dupree and Weinstock was recovered. This result also connects to the discussion in Sect. 6.10.4 on the long time behaviour of a three wave system with diffusion due to turbulence.

References

1. B. B. Kadomtsev, Plasma Turbulence, Academic Press, New York 1965.
2. L.I. Rudakov, Sov. Phys. JETP **21**, 917 (1965).
3. T.H. Dupree, Phys. Fluids **10**, 1049 (1967).
4. T.H. Dupree, Phys. Fluids **9**, 1773 (1966).

5. J. Weinstock, Phys. Fluids **12**, 1045 (1969).
6. B. B. Kadomtsev and O.P. Pogutse, in Reviews of Plasma Physics (Ed. M.A. Leontovitch) Consultant Bureau, New York, Vol. 5, p. 249 (1970).
7. J.B. Taylor and B. McNamara, Phys. Fluids **14**, 1492 (1971).
8. H. Okuda and J.M. Dawson, Phys. Fluids **16**, 408 (1973).
9. R.C. Davidson, Methods in Nonlinear Plasma Theory, Academic Press, New York 1972.
10. G. Joyce, D.C. Montgomery and F. Emery, Phys. Fluids **17**, 110 (1974).
11. A. Hasegawa, Plasma Instabilities and Nonlinear Effects, Springer, New York, 1975.
12. J. Weiland and H. Wilhelmsson, Coherent Nonlinear Interaction of Waves in Plasmas, Pergamon Press, Oxford 1977.
13. H. Sanuki, and G. Schmidt, J. Phys. Soc. Japan **42**, 260 (1977).
14. D. Fyfe and D. Montgomery, Phys. Fluids **22**, 246 (1979).
15. R.Z. Sagdeev, V.D. Shapiro and V.I. Shevchenko, Sov. J. Plasma Phys. **4**, 306 (1978).
16. C.Z. Cheng and H. Okuda, Nuclear Fusion **18**, 87 (1978).
17. W. M. Tang, Nuclear Fusion **18**, 1089 (1978).
18. A. Hasagawa and K. Mima, Phys. Fluids **21**, 87 (1978).
19. C. Chu and M.S. Chu and T. Ohkawa, Phys. Rev. Lett. **41**, 653 (1978).
20. A. Hasagawa, C.G. Maclennan and Y. Kodama, Phys. Fluids **22**, 2122 (1979).
21. A.B. Hassam and R. Kulsrud, Phys. Fluids **22**, 2097 (1979).
22. G. A. Navratil and R.S. Post, Comments on Plasma Physics and Controlled Fusion **5**, 29 (1979).
23. K. Nozaki, T. Taniuti and K. Watanabe, J. Phys. Soc. Japan **46**, 991 (1979).
24. J. Weiland, H. Sanuki, Phys. Lett. **72A**, 23 (1979).
25. V.P. Pavlenko and J. Weiland, Phys. Rev. Lett. **44**, 148 (1980).
26. V.P. Pavlenko and J. Weiland, Phys. Fluids **13**, 408 (1980).
27. J. Weiland, Phys. Rev. Lett. **44**, 1411 (1980).
28. H. Okuda, Phys. Fluids **23**, 498 (1980).
29. A. Hasagawa, H. Okuda and M. Wakatani, Phys. Rev. Lett. **44**, 248 (1980).
30. J. Weiland, H. Sanuki and C.S. Liu, Phys. Fluids **24**, 93 (1981).
31. M.Y. Yu, P.K. Shukla and H.U. Rahman, J. Plasma Phys. **26**, 359 (1981).
32. P.K. Shukla, M.Y. Yu, H.U. Rahman and K.H. Spatschek, Phys. Rev. **A24**, 1112 (1981).
33. K. Katou, J. Phys. Soc. Japan **51**, 996 (1981).
34. J. Weiland, Physica Scripta **23**, 801 (1981).
35. V.P. Pavlenko and J. Weiland, Phys. Rev. Lett. **46**, 246 (1981).
36. R. Nakach, V.P. Pavlenko, J. Weiland and H. Wilhelmsson, Phys. Rev. Lett. **46**, 447 (1981).
37. J. Weiland and J.P. Mondt, Phys. Rev. Lett. **48**, 23 (1982).
38. G. Rogister and G. Hasselberg, Phys. Rev. Lett. **48**, 249 (1982).
39. T. Taniuti, and A. Hasegawa, Physica Scripta **T2:1**, 147 (1982).
40. N. Bekki, H. Takayasu, T. Taniuti and H. Yoshihara, Physica Scripta **T2:2**, 89 (1982).
41. H. Pecseli, Physica Scripta **T2:1**, 83 (1982).
42. D.C. Montgomery, Physica Scripta **T2:1**, 83 (1982).
43. A. Hasagawa and M. Wakatani, Phys. Rev. Lett. **50**, 682 (1983).
44. A. Hasagawa and M. Wakatani, Phys. Fluids **26**, 2770 (1983).
45. R.E. Waltz, Phys. Fluids **26**, 169 (1983).
46. J. Weiland and H. Wilhelmsson, Physica Scripta **28**, 217 (1983).
47. G. Rogister and G. Hasselberg, Phys. Fluids **26**, 1467 (1983).
48. H.U. Rahman and J. Weiland, Phys. Rev. **A28**, 1673 (1983).
49. P. Terry and W. Horton, Phys., Fluids **26**, 106 (1983).
50. H.D. Hazeltine, Phys. Fluids **26**, 3242 (1983).
51. H.L. Pecseli, T. Mikkelsen and S.E. Larsen, Plasma Physics **25**, 1173 (1983).
52. H.L. Pecseli, J. Juul Rasmussen, H. Sugai and K. Thomsen, Plasma Phys. Control. Fusion **26**, 1021 (1984).
53. J. Weiland, Physica Scripta **29**, 234 (1984).

54. P.K. Shukla, M.Y. Yu, H.U. Rahman and K.H. Spatschek, "Nonlinear convective Motion in Plasmas", Physics Reports **105**, 227–328 (1984).
55. J. Weiland and J.P. Mondt, Phys. Fluids **28**, 1735 (1985).
56. P.C. Liewer, Nuclear Fusion **25**, 543 (1985).
57. R.E. Waltz, Phys. Lett. **55**, 1098 (1985).
58. V.I. Petviashvili and O.A. Pokhotelov, JETP Lett. {\bf 42}, 54 (1985).
59. P.K. Shukla, Phys. Rev. **A32**, 1858 (1985).
60. E.A. Witalis, IEE. Trans. Plasma Sci. **14**, 842 (1986).
61. L. Turner, IEE. Trans. Plasma Sci. **14**, 849 (1986).
62. C.T. Hsu, H.D. Hazeltine and J.P. Morrison, Phys. Fluids {\bf 29}, 1480 (1986).
63. T. Taniuti, J. Phys. Soc. Japan **55**, 4253 (1986).
64. D. Jovanovic, H.L. Pecseli, J.J. Rasmussen and K. Thomsen, J. Plasma Physics **37**, 81 (1987).
65. M. Liljeström and J. Weiland, Phys. Fluids **31**, 2228 (1988).
66. Katou and J. Weiland, Phys. Fluids **31**, 2233 (1988).
67. H. Nordman and J. Weiland, Phys. Lett. **A37**, 4044 (1988).
68. A.M. Martins and J.T. Mendoca, Phys. Fluids **31**, 3286 (1988).
69. P.K. Shukla and J. Weiland, Phys. Lett. **A136**, 59 (1989).
70. P.K. Shukla and J. Weiland, Phys. Rev. **A40**, 341 (1989).
71. H. Nordman and J. Weiland, Nuclear Fusion **29**, 251 (1989).
72. B.G. Hong, W. Horton, Phys. Fluids **B2**, 978 (1989).
73. H. Nordman, J. Weiland and A. Jarmen, Nuclear Fusion **30**, 983 (1990).
74. H. Wilhelmsson, Nucl. Phys. **A518**, 84 (1990).
75. M. Persson and H. Nordman, Phys. Rev. Lett. **67**, 3396 (1991).
76. J. Weiland and H. Nordman, Nuclear Fusion **31**, 390 (1991).
77. J. Nycander and V.V. Yankov, Phys. Plasmas **2**, 2874 (1995).
78. H.L. Pecseli and J. Trulsen, J. Plasma Physics **54**, 401 (1995).
79. N. Mattor and S.E. Parker, Phys. Rev. Lett. **79**, 3419 (1997).
80. G.N. Throumoulopoulos and D. Pfirsch, Phys. Rev. **E56**, 5979 (1997).
81. A.Zagorodny and J. Weiland, Phys. Plasmas **6**, 2359 (1999).
82. I. Holod, A. Zagorodny and J. Weiland, Phys. Rev. **E71**, 046401–1 (2005).
83. A. Zagorodny and J. Weiland, Physics of Plasmas. **16** 052308. (2009).

General References

The following references are particularly relevant to the general area covered by this book.

Plasma Physics for Magnetic Fusion

1. N.A. Krall and A.W. Trivelpiece, Principles of Plasma Physics, McGraw-Hill, New York 1973.
2. R.J. Goldston and P.H. Rutherford, An introduction to Plasma Physics, Adam Hilger, Bristol 1995.

Theory of Stability and Transport in Magnetic Confinement Systems

3. W.M. Manheimer and C.N. Lashmore Davies, MHD and Microinstabilities in Confined Plasma, Adam Hilger, Bristol 1989.
4. A. Hasegawa, Plasma Instabilities and Nonlinear Effects, Springer, New York, 1975.
5. B.B. Kadomtsev and O.P. Pogutse in Reviews of Plasma Physics, Consultants Bureau, New York Vol 5, p. 249 (1970).
6. J.W. Connor and H.R. Wilson, Plasma Phys. Control. Fusion **36**, 719 (1994).

J. Weiland, *Stability and Transport in Magnetic Confinement Systems*,
Springer Series on Atomic, Optical, and Plasma Physics 71,
DOI 10.1007/978-1-4614-3743-7,

Nonlinear Effects and Turbulence

7. B.B. Kadomtsev, Plasma Turbulence, Academic Press , New York 1965.
8. R.C. Davidson, Methods in Nonlinear Plasma Theory, Academic Press, New York 1972.
9. J. Weiland and H. Wilhelmsson, Coherent Nonlinear Interaction of Waves in Plasmas, Pergamon Press, Oxford 1977.

Theory and Experiments on Transport

10. P.C. Liewer, Nuclear Fusion **25**, 543 (1985).
11. F. Wagner and U. Stroth, Plasma Phys. Control. Fusion **35**, 1321 (1993).
12. B.A. Carreras, IEEE Trans. Plasma Sci. **25**, 1281 (1997).

Here the citations with titles are books and those without are review papers.\\

Refs. 1 and 2 are comprehensive and include general plasma physics with applications to magnetic fusion. They treat difficult and fundamental problems rigourously and provide excellent basic knowledge for a fusion physicist.

Refs. 3 and 4 are similar to the present book in that they treat both MHD and transport. Ref 4 discusses several instabilities also in the context of space physics and also includes nonlinear effects.

Refs. 5 and 6 are review papers that discuss many instabilities of interest for transport. Ref 6 also presents transport coefficients corresponding to many instabilities.

Ref. 7 is the first and also the most frequently cited book on plasma turbulence.

It is mainly focused towards problems relevant to magnetic fusion and also contains one of the first renormalizations of plasma turbulence.

Ref. 8 is particularly strong on kinetic nonlinear theory. It includes several mathematical tools such as e.g. the method of multiple time scales.

Ref. 9 is more directed towards general plasma physics and Laser Fusion. It does, however, cover problems of nonlinear dynamics and partially coherent wave interactions relevant to the nonlinear saturation of drift wave turbulence.

Refs. 10 and 11 discuss experimental transport research including diagnostics in detail. Ref. 12 is more focused on the relevance of different theories for explaining experimental results.

Answers to Exercises

2.1 $\frac{\delta n}{n} / \frac{e\varphi}{T_e}$

2.2 v_g equals twice the curvature drift after averaging over a Maxwellian distribution.

2.4 The diamagnetic drift is divergence free when grad P is parallel to grad n.

3.1 This is due to the fact that nv_{*} is divergence free, see Eq. 1.4a.

3.2 (a) No difference

3.2 (b) The only difference is that $k_y^2\rho^2$ is replaced by $k_\perp^2\rho^2$.

3.3 These are the effects giving the finite div **A** (compare the discussion following Eq. 1.7). This means that both kinds of ion inertia appearing as $k_y^2\rho^2$ and $k_\|^2c_s^2$ are associated with compressibility.\

3.4 In both cases the inertia term $\omega(\omega - k_y v_{g\,i})$ is replaced by $\omega(\omega-\omega_{*i} - k_y v_{g\,i})$.

3.5 The solution of the dispersion relation can be written $\omega = \omega_r + i\gamma$ where

$$\omega_r = \omega_{*e}(1 - k_y{}^2\rho^2) + k_y v_{gi}$$

$$\gamma = \frac{m_e v_{ei}}{k_\|{}^2 T_e}\omega_{*e}\left[k_y{}^2\rho^2\omega_{*e} + k_y(v_{ge} - v_{gi})\right]$$

3.6 $\beta < m^2 / \frac{d\ln P}{dr}$

3.7 $\beta_c = \left(\frac{a}{qR}\right)^2$

3.8 The intermediate result is

$$\frac{\delta n_e}{n} = \frac{\omega_{*e}}{\omega}\frac{e\phi}{T_e} - k^2\rho^2\frac{k_\|{}^2 v_A{}^2}{\omega^2}\frac{\omega}{k_\|}\frac{eA_\|}{T_e}$$

3.9 $E_\| = -ik_\|\phi\dfrac{k^2\rho^2 k_\|{}^2 v_A{}^2}{\omega(\omega_{*e} - \omega) + k^2\rho^2 k_\|{}^2 v_A{}^2 + k_\|{}^2 c_s{}^2}$

3.10 $m \approx 280$ for $q = 2$

3.11 $\omega^2(1 + k_y{}^2\rho^2) - \omega\omega_{*e}(1 - k_y{}^2\rho^2) - k_\|{}^2 c_s{}^2 = 0$

J. Weiland, *Stability and Transport in Magnetic Confinement Systems*,
Springer Series on Atomic, Optical, and Plasma Physics 71,
DOI 10.1007/978-1-4614-3743-7,

Typical Parameter Values for a Tokamak Plasma

We will here give numerical values of some of the most important quantities associated with low frequency modes in tokamak plasma. The basic machine performance is taken from JET.

With magnetic field of
B = 27.7 10^3 Gauss = 2.77 T
we find the cyclotron frequencies
$\Omega_{ci} = 2.65\ 10^8\ s^{-1}$
$\Omega_{ce} = 4.87\ 10^{11}\ s^{-1}$
A density of
$n = 10^{20}\ m^{-3}$
corresponds to the plasma frequencies
$\omega_{pe} = 5.6\ 10^{11}\ s^{-1}$
$\omega_{pi} = 1.3\ 10^{10}\ s^{-1}$
and the Alfvén velocity
$v_A = B/(\mu_0 n m_i)^{1/2} = 0.6\ 10^7$ m/s
This gives the dielectric constant for flute modes ($k_\| = 0$)

$$\begin{aligned}\varepsilon &= 1 + \mu_0 \rho_m c^2/B^2 = 1 + c^2/{v_A}^2 = 1 + {\omega_{pi}}^2/{\Omega_{ci}}^2 \\ &\approx 1 + (\rho/\lambda_{de})^2 \approx 2406(\rho/\lambda_{de} = 49)\end{aligned}$$

where ρ_m is the mass density and $\rho = c_s/\Omega_{ci}$.

We also notice that
$\omega_{pe} = 1.15\ \Omega_{ce}$
At fusion temperatures
$T_e = T_i = 10^8$ K = 8.6 keV = $1.38\ 10^{-15}$ J
We find

$\lambda_{de} = (\varepsilon_0 T_e/ne^2)^{1/2} = 0.69\ 10^{-2}$ cm
$g = (n\ {\lambda_{de}}^3)^{-1} = 3\ 10^{-8}$
$v_{the} = (2T_e/m_e)^{1/2} = 0.55\ 10^8$ m/s
$v_{thi} = (2T_i/m_i)^{1/2} = 1.29\ 10^6$ m/s

$\rho_e = v_{the}/\Omega_{ce} = 1.13\ 10^{-2}$ cm
$\rho_i = v_{thi}/\Omega_{ci} = 0.49$ cm

$\nu_{ei} = 0.5\ 10^4\ s^{-1}$
$D_e = \nu_{ei}\ {\rho_e}^2 = 0.6\ 10^{-4}\ m^2/s$

With a major radius
R = 3 m
and a minor radius
a = 1 m
we find
$\kappa = 1/a = 1\ m^{-1}$

$v_{*e} = \kappa T_e/(eB) = 3.2\ 10^3$ m/s
$v_g = g/\Omega_{ci} = (T_i/m_i)/(r\ \Omega_{ci}) = 1.1\ 10^3$ m/s
$\omega_{int} = (\kappa g)^{1/2} = 5\ 10^5\ s^{-1}$

For $q = 2$ and $k_\| = 1/qR$ we have
$k_\| v_A = 10^5\ s^{-1}$
with a plasma current
$I = 2.6\ 10^6$ A
we have the average electron current velocity
$v_{be} = 0.5\ 10^5$ m/s

Index

J. Weiland, *Stability and Transport in Magnetic Confinement Systems*,
Springer Series on Atomic, Optical, and Plasma Physics 71,
DOI 10.1007/978-1-4614-3743-7,

MIX
Papier aus verantwortungsvollen Quellen
Paper from responsible sources
FSC® C105338

If you have any concerns about our products,
you can contact us on
ProductSafety@springernature.com

In case Publisher is established outside the EU,
the EU authorized representative is:
Springer Nature Customer Service Center GmbH
Europaplatz 3, 69115 Heidelberg, Germany

Printed by Libri Plureos GmbH
in Hamburg, Germany